Product Design

teNeues

Product
Design

teNeues

Editor in chief:	Paco Asensio
Editor and texts:	Héctor Roqueta
Editorial coordination:	Cynthia Reschke
Research:	Marta Casado
Art director:	Mireia Casanovas Soley
Layout:	Ignasi Gràcia Blanco
Copy-editing:	Francesc Bombí-Vilaseca, Simone K. Schleifer
German translation:	Inken Wolthaus
French translation:	Leïla Marçot
English translation:	Matthew Clarke

Published by teNeues Publishing Group

teNeues Publishing Company
16 West 22nd Street, New York, NY 10010, US
Tel.: 001-212-627-9090, Fax: 001-212-627-9511

teNeues Book Division
Neuer Zollhof 1
40221 Düsseldorf, Germany
Tel.: 0049-(0)211-994597-0, Fax: 0049-(0)211-994597-40

teNeues Publishing UK Ltd.
Aldwych House, 71/91 Aldwych
London WC2B 4HN, UK

www.teneues.com

| ISBN: | 3-8238-5597-2 |

| Editorial project: | © 2002 LOFT Publications |

Domènech 9, 2-2
08012 Barcelona, Spain
Tel.: 0034 932 183 099
Fax: 0034 932 370 060

e-mail: loft@loftpublications.com
www.loftpublications.com

| Printed by: | Gràfiques Anman del Vallès. Barberà del Vallès. Spain |

November 2002

Bibliographic information published by Die Deutsche Bibliothek Die Deutsche Bibliothek lists this publication in the Deutsche Nationalbibliographie; detailed bibliographic data is available in the Internet at http://dnb.ddb.de.

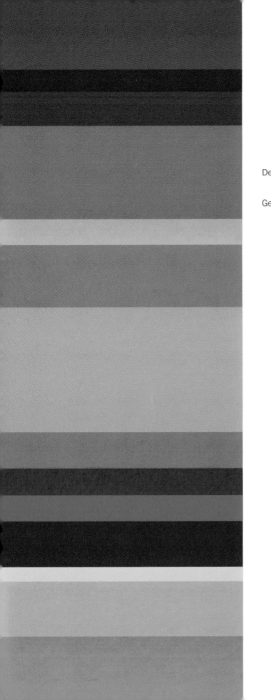

Einführung

Introduction

Introduction

Introducción

Introduction

All the objects surrounding us have been designed—this is something intrinsic to our society—but not all of them have been well designed. A well-designed product is not only characterized by its great practical utility; it also has to convey something more, in order to be attractive and so distinguish itself from the competition. This book provides a selection of well-designed products from the last few years, created with the aim of making our everyday life both more attractive and more convenient.

The beginning of the 21st century marks a good point from which to survey the current state of industrial design, as well as considering the role of the industrial designer in the new millennium.

It is clear that we live in the beginning of the age of telecommunications. The new technologies are increasingly present in our lives, and, therefore, in industrial design, where CAD/CAM systems, Rapid Prototyping and 3-D images make the design process easier and more flexible. However, we must not be blinded by computer science, which is a mere tool; designers also need to use technology to look for new materials and come up with solutions for the accumulation of waste products, new social practices and our increasing dependence on objects.

A product's design has to take into account technological innovations, economic conditions, society's changing priorities and the latest developments in art and architecture. All these factors influence design and give rise to constant changes. Nowadays, the latest technology is reducing the size of objects, while globalization and the opening up of international markets mean that a single product can be aimed at people from widely different cultural backgrounds. Should we create a new global culture for objects or should we use them to assert different, local identities?

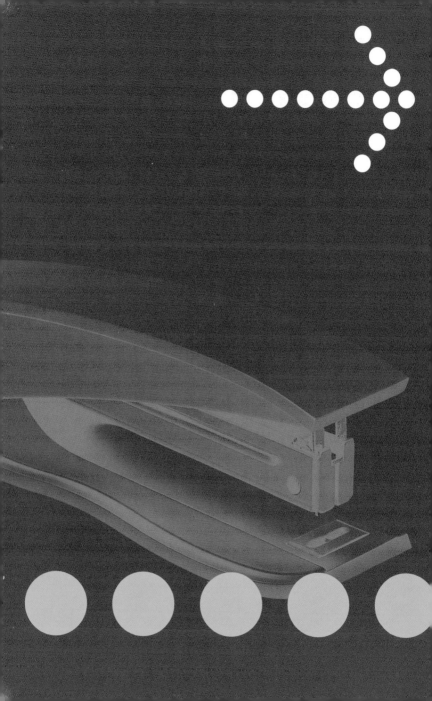

Einführung

Es liegt im Wesen unserer heutigen Gesellschaft, dass alle uns umgebenden Objekte einem Design unterworfen sind, jedoch sind nicht alle Objekte von einem guten Design bestimmt. Ein Produkt mit einem guten Design zeichnet sich nicht nur durch seine große praktische Nützlichkeit aus, sondern muss außerdem einen gewissen Aussagewert haben, um attraktiv zu erscheinen und sich von den restlichen Objekten der Konkurrenz zu unterscheiden. In diesem Buch wird eine Auswahl von Produkten der letzten Jahre mit einem guten Design gezeigt, die unser tägliches Leben attraktiver und gleichzeitig leichter gestalten.

Der Beginn des 21. Jahrhunderts ist ein guter Moment, um das Panorama des industriellen Designs zu betrachten und über die Rolle des Industriedesigners in diesem Jahrhundert nachzudenken.

Wir stehen ohne Zweifel am Beginn des Zeitalters der Telekommunikationen. Neue Technologien nehmen einen immer größeren Platz in unserem Umfeld und im industriellen Design ein, wo CAD/CAM-Systeme, das Rapid Prototyping oder 3D-Bilder den Designprozess erleichtern und flexibler gestalten. Wir dürfen uns jedoch nicht nur auf die Anwendungsmöglichkeiten des Computers konzentrieren, der letztlich nur ein Werkzeug ist, sondern die Technologie dazu benutzen, neue Materialien zu entdecken sowie Problemlösungen beispielsweise für die massive Anhäufung von Abfällen, neue gesellschaftliche Formeln oder die immer enger werdende Beziehung zu den Objekten zu entwickeln.

Das Design eines Produktes entsteht auf dem halben Weg zwischen den technologischen Neuerungen, den wirtschaftlichen Umständen, den Wechseln der gesellschaftlichen Bedürfnisse und der Entwicklung in Kunst und Architektur. Alle diese Faktoren beeinflussen und bewirken eine konstante Veränderung des Designs, so dass einerseits die Technologie der letzten Generation die Objekte entmaterialisiert und andererseits die Verallgemeinerung und die globale Öffnung der Märkte die Erschaffung eines einzigen Produktes für Menschen verschiedener kultureller Wurzeln zulässt. Müssen wir eine neue globale Kultur des Objektes schaffen oder es im Gegenteil dazu benutzen, Unterschiede aufzuheben?

Introduction

Tous les objets qui nous entourent ont été conçus, c'est une qualité intrinsèque de notre société. Mais tous les objets ne sont pas nécessairement bien conçus. Un objet bien conçu ne se définit pas uniquement en fonction de sa grande utilité pratique. Il doit également communiquer ou expliquer ce quelque chose supplémentaire qui le rend séduisant et le différencie ainsi des autres objets de la concurrence. Ce livre présente une sélection de quelques produits bien conçus, des années récentes, qui rendent notre vie quotidienne plus simple et agréable.

Les débuts du XXIème sont des instants choisis pour porter un regard sur le panorama actuel du design industriel et lancer une réflexion quant au rôle du créateur industriel dans ce siècle.

Nous sommes incontestablement à l'aube de l'ère des télécommunications. Les nouvelles technologies se font chaque jour plus présentes dans notre environnement comme dans le design industriel, où les systèmes de CAO/DAO, le prototypage rapide voire l'imagerie 3D facilitent et fluidifient le processus de conception. Mais nous ne devons pas nous laisser leurrer par ces seules applications informatiques, qui ne laissent pas d'être des outils, et mettre la technologie au service de la quête de nouveaux matériaux afin de générer de nouvelles solutions à des questions comme la massification des déchets, les nouveaux actes sociaux ou la relation, chaque jour plus étroite, à l'objet.

La conception d'un produit trouve son origine à mi-chemin entre les innovations technologiques, les conditions économiques, les modifications des besoins sociétaux et l'évolution de l'art et de l'architecture. L'ensemble de ces facteurs influe et provoque la constante transformation du design. Ainsi, d'un côté, la technologie de dernière génération dématérialise les objets, et d'un autre côté, la mondialisation et l'ouverture globale des marchés supposent de projeter un même produit pour des personnes aux substrats culturels différents. Devons-nous créer une nouvelle culture globale de l'objet ou, au contraire, en profiter pour diffuser les différences?

Introducción

Todos los objetos que nos rodean están diseñados –es una cualidad intrínseca en nuestra sociedad–, pero no todos los objetos están bien diseñados. Un producto bien diseñado no solamente se define por su gran utilidad práctica, sino que además ha de saber comunicar algo, ha de explicar algo más para hacerse atractivo y poder así diferenciarse de los demás objetos de la competencia. Este libro es una selección de productos bien diseñados en los últimos años que nos facilitan y hacen más atractiva nuestra vida cotidiana.

El inicio del siglo XXI es un buen momento para observar el panorama actual del diseño industrial y así poder reflexionar sobre cuál ha de ser el papel del diseñador industrial en este siglo.

No hay duda que estamos en el principio de la era de las telecomunicaciones. Las nuevas tecnologías están cada vez más presentes en nuestro entorno, también en el diseño industrial, donde los sistemas CAD/CAM, el Rapid Prototyping o las imágenes en 3D facilitan y agilizan el proceso de diseño. Sin embargo, estas aplicaciones de la informática, que no dejan de ser herramientas, no deberían cegarnos, y deberíamos aprovechar la tecnología en busca de nuevos materiales para nuevas soluciones a problemas como la masificación de residuos, los nuevos actos sociales o la cada vez más estrecha relación con los objetos.

El diseño de un producto se origina a medio camino entre las innovaciones tecnológicas, las circunstancias económicas, los cambios en las necesidades de la sociedad y la evolución del arte y la arquitectura. Todos estos factores influyen y provocan la constante transformación en el diseño; así, por un lado la tecnología de última generación está desmaterializando los objetos, y, por otro, la mundialización y la apertura global de los mercados suponen proyectar un mismo producto para gente con diferentes raíces culturales. ¿Debemos crear una nueva cultura global del objeto o por el contrario aprovechar para difundir las diferencias?

History

Geschichte

Histoire

Historia

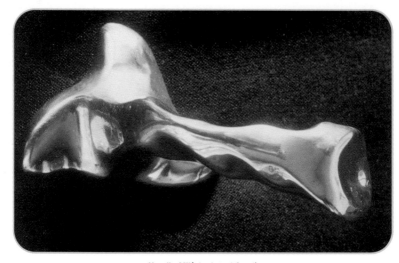

Handle Milà *by Antoni Gaudí*

ARTS & CRAFTS

Industrial design as we now know it appeared in the middle of the 19th century as a consequence of the division of labor imposed by the Industrial Revolution. A single craftsman was no longer responsible for both drawing up an object and making it a reality; the growing specialization in the workplace meant that somebody had to produce a design that would serve as a basis for the manufacturing process. The Arts & Crafts movement was the first to become aware of not only the designer's function but also the social consequences involved in industrialization.

Ideologically, the movement leaned towards an utopian socialism that rejected machines on account of their alienating effects and defended craftsmanship and the trade guilds as means of dignifying work.

It was a movement that sought not only social reform but also stylistic innovation. It rejected the pretentious and trivial decorations of the era's manufactured products and advocated a beauty that was accessible to all. They sought to respect the intrinsic qualities of the materials they used and based their decorations on forms found in nature.

The Arts & Crafts movement represented sensitivity in the face of industrialization; its achievements were later consolidated in Europe by Art Nouveau in France, the Jugendstil in Germany, the Modern Style in England and, above all, by Modernism in Spain.

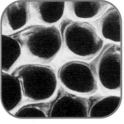

Ironwork Calvet
by Antoni Gaudí

ARTS & CRAFTS

Das industrielle Design, so wie wir es heute verstehen, erscheint zum ersten Mal Mitte des 19. Jahrhunderts im Zuge der von der industriellen Revolution eingeführten Arbeitsteilung. Heute ist nicht mehr das Individuum verantwortlich für den Entwurf und die Ausführung eines Objektes; angesichts der zunehmenden Spezialisierung der Arbeit muss vor Beginn der Durchführungsphase ein Schema erstellt werden.

Arts & Crafts ist die erste Bewegung, die die Aufgabe des Designers erkannt hat und sich der gesellschaftlichen Konsequenzen der Industrialisierung bewusst geworden ist.

Ihre Ideologie basierte auf einem utopischen Sozialismus, der die Maschine mit ihrem entfremdenden Einfluss ablehnte und das Kunstgewerbe und dessen genossenschaftliche Organisation zur Würdigung der Arbeit verteidigte.

Diese Bewegung beabsichtigte gleichzeitig eine gesellschaftliche Umstrukturierung sowie eine Erneuerung des Stiles. Die anmaßende und oberflächliche Dekoration der Objekte dieser Zeit wurde abgelehnt und die für alle zugängliche Schönheit gesucht. Dazu blieb man weiterhin den verwendeten Materialien treu und passte ihre Dekoration den Formen der Natur an.

Gegenüber der Industrie bedeutete dies eine Sensibilisierung, die in Europa später als Art Nouveau in Frankreich, als Jugendstil in Deutschland, Modern Style in England und ganz besonders im spanischen Modernismo Form annahm.

Chair Calvet *by Antoni Gaudí*   **Bench Batlló** *by Antoni Gaudí*

ARTS & CRAFTS

Le design industriel, comme nous l'entendons aujourd'hui, apparaît au milieu du XIXème siècle en conséquence de la division du travail imposée par la révolution industrielle. Nous ne sommes plus confrontés à un artisan responsable de projeter et de créer un objet, mais bien à la spécialisation croissante du travail requérant que quelqu'un réalise un plan préalable à toute fabrication.

L'Arts & Crafts est le premier mouvement qui prend conscience de la fonction du concepteur, ainsi que des conséquences sociales inhérentes à l'industrialisation.

Idéologiquement, le mouvement repose sur un socialisme utopique qui rejette la machine en raison de ses effets aliénants et défend l'artisanat et son organisation corporatiste afin de rendre sa dignité au travail.

Le mouvement constituait non seulement une tentative de réforme sociale mais aussi d'innovation stylistique. Dans cette optique, la décoration prétentieuse et superficielle des objets fabriqués à l'époque était écartée en faveur d'une beauté à portée de tous. Pour ce faire, le mouvement demeurait fidèle aux matériaux utilisés et fondait sa décoration sur les formes du monde naturel.

Un instant de sensibilisation face à l'industrie, consolidé par la suite dans toute l'Europe avec l'Art Nouveau en France, le Jugendstil en Allemagne, le Modern Style en Angleterre et surtout le Modernisme en Espagne.

Bench Calvet *by Antoni Gaudí*

ARTS & CRAFTS

El diseño industrial tal y como hoy lo entendemos aparece
a mediados del siglo XIX como consecuencia de la división
del trabajo impuesta por la revolución industrial. Ya no estamos ante un artesano responsable de proyectar y ejecu-
tar un objeto, sino que ante la creciente especialización del
trabajo se reclama que alguien realice un esquema previo
a cualquier fabricación.

El Arts & Crafts es el primer movimiento que toma con-
ciencia de la función del diseñador, así como de las conse-
cuencias sociales que conlleva la industrialización.

Ideológicamente sus seguidores se basaron en un socialis-
mo utópico que rechazaba la máquina por su efecto alie-
nante y defendía la artesanía y su organización gremial para dignificar el trabajo.

Fue un movimiento basado en un intento de reforma social
pero también de innovación de estilo. Rechazó la decoración pretenciosa y superficial de los objetos fabricados de
la época y buscó la belleza al alcance de todos. Para ello
se guardó fidelidad a los materiales utilizados y se basó la
decoración en las formas del mundo natural.

Fue un punto de sensibilización frente a la industria que se
consolidó posteriormente en Europa con el Art Nouveau en
Francia, el Jugendstil en Alemania, el Modern Style en In-
glaterra y, sobre todo, el Modernismo en España.

BAUHAUS

What was the Bauhaus? It was not an artistic movement or a modern style. The Bauhaus was a school that opened in Germany in 1919, and in the fourteen years of its existence it provided the basis for a new profession: industrial design. An industrial designer has to master and apply art and technology in equal measures. So, the Bauhaus created a teaching program that brought the workshop into the school, thereby creating the model for most of today's design schools all over the world.

The designer faces three main challenges: the first is to discover the true essence of an object, what it is that really defines it; the second is to study its function, to be able to improve it; and the final one is to search for its beauty.

The study of geometry and elemental forms, as well as the influence of Constructivism and, above all, De Stijl, gave rise to an easily identifiable formalist style in the Bauhaus, based on cubes, cones, spheres and cylinders. It established the canons of modern furniture by experimenting with bent tubes of stainless steel.

The Bauhaus sought to bring art to the people by integrating it into the industrial process but, paradoxically, only an intellectual elite understood its work.

Chronograph watch
*for Projects by Michael Graves
& Associates*

BAUHAUS

Was ist das Bauhaus? Es ist weder eine künstlerische Bewegung noch ein moderner Stil. Das Bauhaus ist eine 1919 gegründete Schule, die während der vierzehn Jahre ihres Bestehens die Grundlagen für einen neuen Beruf legte: den industriellen Designer.

Dieser musste in seiner beruflichen Praxis sowohl die Kunst als auch die Technik beherrschen und anwenden können. Dazu wurde ein pädagogisches Programm ausgearbeitet, das in die Schule eine Werkstatt integrierte, und das noch heute bei den meisten Designer-Schulen der Welt Modell steht.

Der Designer musste mehreren Zielsetzungen gerecht werden: zuerst musste er das wirkliche Wesen des Objektes entdecken, das es definiert. Danach kam die Untersuchung seiner Funktion, um eventuelle Verbesserungen vorzunehmen und zum Schluss blieb die Suche nach seiner Schönheit.

Die Vorliebe für die Geometrie und die elementaren Formen sowie der Einfluss des Konstruktivismus und vor allem von De Stijl schufen einen gewissen formalistischen, leicht zu identifizierenden Stil des Bauhauses, der auf Würfeln, Kegeln, Kugeln und Zylindern basierte. Es wurde mit gefalteten Stahlrohren experimentiert und die Richtlinien des modernen Mobiliars festgelegt.

Durch die Integration der Kunst in die industriellen Prozesse sollte sie dem Volk nahe gebracht werden; paradoxerweise wurden diese Produkte jedoch nur von einer intellektuellen Elite verstanden.

BAUHAUS

Qu'est le Bauhaus? Ce n'est ni un courant artistique, ni un style moderne. Le Bauhaus était une école allemande créée en 1919 qui durant ses quatorze années d'existence assit les bases d'une nouvelle profession : le design industriel.

Le designer, dans sa pratique, devait dominer et appliquer de la même manière l'art et la technique. Pour ce faire, fut instauré un programme pédagogique introduisant l'atelier dans l'école, un exemple encore de nos jours repris dans la plupart des écoles de design de par le monde.

Le designer avait plusieurs objectifs : le premier était de découvrir l'essence réelle de l'objet, ce qui le définit. Le deuxième était d'étudier sa fonction afin d'être à même de l'améliorer. Il devait, enfin, rechercher sa beauté.

L'étude de la géométrie et des formes élémentaires, de même que l'influence du constructivisme et surtout du mouvement De Stijl, engendra dans la Bauhaus un certain style formaliste reposant sur des cubes, cônes, sphères et cylindres aisément identifiables. Ainsi furent établis les canons du mobilier moderne expérimentés à l'aide de tubes d'acier pliés.

La prétention était de rapprocher l'art du peuple en l'intégrant à l'industrie. Paradoxalement, seule une élite intellectuelle fut à même de comprendre ces produits.

BAUHAUS

¿Qué es la Bauhaus? No es una corriente artística ni un estilo moderno. Bauhaus fue una escuela alemana creada en 1919 que en sus catorce años de existencia sentó las bases de un nuevo profesional: el diseñador industrial.

Éste, en su práctica profesional, debía dominar y aplicar por igual el arte y la técnica. Para ello se creó un programa pedagógico que introdujo el taller dentro la escuela, algo que aún hoy sirve de base en la mayoría de escuelas de diseño del mundo.

El diseñador tenía varios objetivos: el primero era descubrir la verdadera esencia del objeto, aquello que lo define. El segundo era el estudio de su función, para poder mejorarla y por último debía buscar su belleza.

El estudio de la geometría y de las formas elementales, así como la influencia del constructivismo y sobretodo del De Stijl, creó en la Bauhaus un cierto estilo formalista fácilmente identificable basado en los cubos, conos, esferas y cilindros. Estableció los cánones del mobiliario moderno experimentando con tubos de acero doblado.

La Bauhaus pretendió acercar el arte al pueblo mediante la integración de éste en la industria. Paradójicamente, tan solo una elite intelectual comprendió sus productos.

GOOD DESIGN

In the mid-1950s, coinciding with the resurgence of German industry, the design world developed a formal language dominated by the watchwords: practical, economical and rational. This quickly spread throughout Europe and became the official doctrine for decades to come.

"Forms follows Function" was the order of the day. The designer's job was to analyze an object in order to eliminate any superfluous features and so attain the maximum degree of functionality.

The idea was to make the product as useful as possible by achieving an optimal performance and ensuring absolute clarity for the user. Ergonomic questions were taken into account and the design was meticulously pondered down to the very last detail.

Braun, the manufacturer of small, domestic electrical apparatuses associated with the Design College of Ulm, became the standard bearer for this new functionalism.

GOOD DESIGN

Der industrielle Aufschwung Deutschlands Mitte der fünfziger Jahre fiel zusammen mit der Entwicklung einer formalen Sprache in der Welt des Designs, die von Begriffen beherrscht wurde wie praktisch, wirtschaftlich und rational, und die sich in kurzer Zeit in ganz Europa als offizielle Doktrin während mehrerer Jahrzehnte durchsetzte.

„Form follows Function" waren die herrschenden Maxime und Axiome. Die Aufgabe des Designers war, das Objekt zu analysieren und es von allen oberflächlichen Attributen zu befreien, um so seine optimale Funktionalität zu gewährleisten.

Man suchte die maximale Nützlichkeit des Produktes, seine optimale und für den Anwender eindeutige Funktionsweise. Ergonomische Gesichtspunkte wurden berücksichtigt und das Design bis hin zu den kleinsten Details sorgfältig ausgearbeitet.

Als die für ihre elektrischen Haushaltsgeräte bekannte Firma Braun sich mit der Hochschule für Design in Ulm zusammenschloss, wurden sie zu einem Musterbeispiel dieses neuen Funktionalismus.

Chair by Le Corbusier

Chair *by Mies van der Rohe*

GOOD DESIGN

Au milieu des années 50, coïncidant avec la renaissance industrielle de l'Allemagne, un langage formel se développa dans le monde du design dominé par des concepts comme : pratique, économique et rationnel. Rapidement, il devait s'étendre à toute l'Europe pour devenir la doctrine officielle durant plusieurs décennies.

« Form follows Function – La forme suit la fonction » , tels étaient la maxime et l'axiome à suivre. La fonction du designer portait sur l'analyse de l'objet, afin de le dévêtir de toute superficialité et obtenir ainsi une fonctionnalité maximale.

La recherche portait sur la maximisation de l'utilité du produit afin que son fonctionnement devienne optimal et clair pour l'utilisateur. Avant tout étaient pris en considération les aspects ergonomiques et le design était soigné jusque dans les moindres détails.

Associées, l'entreprise de petit électroménager Braun et l'École Supérieure de Design d'Ulm devinrent le paradigme de ce nouveau fonctionnalisme.

GOOD DESIGN

A mediados de los años 50, coincidiendo con el resurgir industrial de Alemania, se desarrolló en el mundo del diseño un lenguaje formal dominado por conceptos como: práctico, económico y racional. Rápidamente se extendió por toda Europa y se convirtió en la doctrina oficial durante varias décadas.

La máxima y el axioma a seguir era: "La forma sigue la función". La función del diseñador era el análisis del objeto para despojarlo de cualquier superficialidad y así conseguir el máximo grado de funcionalidad. Se buscaba la máxima utilidad del producto, que su funcionamiento fuera óptimo y claro para el usuario. Se tenían en consideración los aspectos ergonómicos y se cuidaba el diseño hasta en los detalles más pequeños.

La empresa de pequeños electrodomésticos Braun y su asociación con la Escuela Superior de Diseño de Ulm se convirtieron en el paradigma de este nuevo funcionalismo.

Hot Bertaa
by Philippe Starck

MOVEMENTS IN REACTION

At the end of the 1960s countercultural movements appeared in both politics and art to question and attempt to change the established order.

In the design field, various practitioners formed small, loose groups, particularly in Germany and Italy, to create Utopias and conceptual designs in keeping with the spirit of the age.

A decade later a new wave of designers abandoned these Utopian concepts but renewed the assault on the Calvinist doctrine of functionalism. The prevailing Good Design was too neutral and boring, and it was leading people along a path to a dangerously conformist consumer society.

As a reaction, objects marked by a great intellectual and formal irreverence began to appear, mixing the most diverse materials with Pop iconography to search for new forms of expression. The color and sense of humor of these designs quickly attracted the attention of the media.

In fact, most of these objects were extremely expensive pieces of furniture that became icons for those who owned them. However, they did manage to change the relationship between Man and object by placing as much, or even more, emphasis on the latter's symbolic function than its practical one.

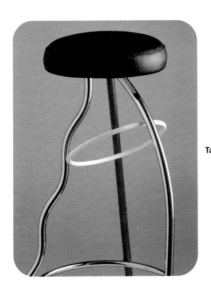

Taburette Duplex
by Mariscal

REAKTIONÄRE BEWEGUNGEN

Gegen Ende der sechziger Jahre entstanden sowohl politisch als auch kulturell reaktionäre Bewegungen, die die eingeführten Normen in Frage stellten und einen Wechsel anstrebten.

In der Welt des Designs bildeten sich mehr oder weniger kurzlebige Gruppen, deren Ziel es war, Utopien und konzeptionelle Designs im Einklang mit dem Zeitgeist zu schaffen.

Ein Jahrzehnt später wandte sich eine neue Gruppe von Designern von den utopischen Konzepten ab, opponierte aber weiterhin gegen die kalvinistische Doktrin des Funktionalismus. Das herrschende „Good Design" war zu neutral und zu langweilig und führte schließlich zu einer gefährlich einförmigen Konsumgesellschaft frei. Als Reaktion darauf werden ohne Rücksicht auf intellektuelle oder formale Normen Produkte entwickelt, die auf der Suche nach neuen Ausdrucksmöglichkeiten die verschiedenartigsten Materialien mit Pop-Strukturen vermischen. Die Farbgestaltung und der Sinn für Humor dieser Objekte fand über die Kommunikationsmedien in kurzer Zeit weite Verbreitung.

Bei den meisten Objekten handelte es sich um sehr kostbare Möbelstücke, die von ihren Besitzern wie Ikonen geschätzt wurden. Hierdurch erfuhr die Beziehung zwischen Mensch und Objekt jedoch eine Veränderung, bei der der symbolischen Funktion größere Bedeutung zugeschrieben wurde als der praktischen Aufgabe.

A television built from pieces of wood, pressed and glued without formaldehyde.

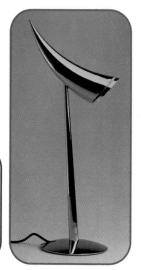

Ara *by Philippe Starck*

MOUVEMENTS DE RÉACTION

La fin des années 60 vit apparaître les mouvements de contre-culture, tant politiques qu'artistiques, mettant en question et essayant de modifier les normes établies.

Dans le monde du design, divers professionnels, surtout allemands et italiens, s'unirent en petits groupes plus ou moins éphémères afin de créer des utopies et des designs conceptuels, en résonance avec l'esprit de l'époque.

Une décennie plus tard, une nouvelle génération de créateurs abandonnait les concepts utopiques, tout en maintenant l'opposition à la doctrine calviniste du fonctionnalisme. L'empire du « Good Design » était par trop neutre et ennuyeux et nous conduisait tous vers une société de consommation dangereusement homogène.

En réaction apparaissaient des objets d'une grande irrévérence intellectuelle et formelle, qui peuvent mêler les matériaux les plus divers à l'iconographie pop, en quête d'une nouvelle force expressive. La couleur et le sens de l'humour des objets se diffusa rapidement par le biais des médias.

De fait, la majeure partie des objets étaient des pièces de mobilier relativement onéreuses, se convertissant en icônes pour leur détenteur. Elles permirent, cependant, de modifier la relation de l'homme à l'objet en conférant une importance égale, voire supérieure, à la fonction symbolique plutôt qu'à la pratique.

Nutty the Cracker *for Alessi by Stefano Giovannoni*

MOVIMIENTOS DE REACCIÓN

A finales de los sesenta aparecieron movimientos contra-culturales, tanto en política como en arte, que cuestiona-ron e intentaron cambiar las normas establecidas.

En el mundo del diseño, diversos profesionales, sobre todo alemanes e italianos, se unen en pequeños grupos más o menos efímeros para crear utopías y diseños conceptuales acordes con el espíritu de la época.

Una década más tarde una nueva hornada de diseñadores abandonó los conceptos utópicos, pero continuó oponién-dose a la doctrina calvinista del funcionalismo. El Good Design imperante era demasiado neutro y aburrido, nos conducía a todos hacia una sociedad de consumo peligro-samente homogénea.

Frente a ello aparecen objetos de una gran irreverencia in-telectual y formal, donde está permitido mezclar los mate-riales más dispersos con la iconografía pop en busca de una nueva expresividad. El colorido y el sentido del humor de los objetos fue rápidamente difundido por los medios de comunicación.

De hecho la mayoría de los objetos fueron piezas de mobi-liario muy caras que se convirtieron en iconos para quien las poseía. De esta forma consiguieron cambiar la relación entre hombre y objeto dándole igual o más importancia a la función simbólica que a la práctica.

Montages *by Lexon*

REDEFINITION OF FORM

The chip or microprocessor is increasingly making its presence felt in the objects we use everyday, and great efforts are being made to extend its use even further. The chip has obvious advantages: it can enhance a product's performance while reducing its size to a minimum, as in the case of the so-called latest-generation objects, which are continually increasing their functions but at the same time get smaller and smaller.

This reduction in the size of so many objects is one of the great challenges confronting industrial design. The maxim "Form follows Function" has become obsolete in the face of this new technology. The organization and functioning of a microprocessor are not perceptible to the senses. The technical function is concentrated in one tiny "box", giving the designer greater freedom in the final configuration of an object.

With all this formal freedom at their disposal, today's designers have to take advantage of multidisciplinary work practices to strengthen the human element in the technological process and change the relationship between Man and objects.

NEUDEFINIERUNG DER FORM

Der Chip oder Mikroprozessor setzt seine Anwesenheit bei den täglichen Gegenständen immer mehr durch und sein Einsatz spielt eine immer größere Rolle. Die Vorteile des Chip sind eindeutig: er vervielfacht die Leistungen des Produktes bei gleichzeitiger Reduzierung seines Volumens auf ein Minimum. Denken wir an die Objekte der sogenannten letzten Generation, deren Funktionen immer vielfältiger und deren Maße immer kleiner werden.

Eine der großen Herausforderungen des industriellen Designs ist die Entmaterialisierung der Objekte. Das Musterbeispiel des „Form follows Function" erscheint im Vergleich mit dieser neuen Technologie überholt. Die Organisation und Funktionsweise eines Mikroprozessors können wir mit unseren Sinnen nicht wahrnehmen. Die technische Funktion wird auf ein kleines „Kästchen" reduziert, so dass der Designer für die endgültige Gestaltung des Objektes über wesentlich mehr Freiheit verfügt.

Angesichts dieser formalen Freiheit sollte der Designer mit Hilfe der vielfältigen modernen Arbeitsdisziplinen versuchen, die Technik zu humanisieren und die Beziehung zwischen Mensch und Objekt zu verändern.

Design by Lexon

REDÉFINITION DE FORME

La puce, ou microprocesseur, est toujours plus présente dans les objets de notre quotidien et tout est mis en œuvre pour que son implantation soit encore plus importante. Les avantages de la puce sont évidents : elle permet d'augmenter les prestations du produit en réduisant son volume au minimum. Ainsi les objets dits de « dernière génération » qui ne cessent d'augmenter leurs capacités tout en réduisant considérablement leur encombrement.

L'un des grands défis du design industriel porte sur la dématérialisation qui frappe nombre d'objets. Le paradigme du « Form follows Function – La forme suit la fonction » devient obsolète face à cette nouvelle vague technologique. L'organisation et le fonctionnement d'un microprocesseur sont imperceptibles à travers nos sens. Tout se réduit à une petite « boîte » qui abrite la fonction technique et, conséquemment, le concepteur possède une liberté plus ample pour déterminer la forme finale de l'objet.

Devant ce surcroît de liberté formelle, le designer devrait, grâce à la multi-disciplinarité actuelle du travail, rendre la technique plus humaine et changer la relation homme / objet.

LA REDEFINICIÓN DE LA FORMA

El chip o microprocesador está cada vez más presente en los objetos que utilizamos diariamente y se está trabajando para que su implantación sea cada vez mayor. Las ventajas del chip son evidentes: permite aumentar las prestaciones del producto reduciendo su volumen al mínimo como los llamados objetos de última generación, que no cesan de aumentar sus funciones reduciendo considerablemente su cuerpo.

Uno de los grandes retos del diseño industrial es la desmaterialización que están sufriendo muchos objetos. El paradigma de "la forma sigue la función" queda obsoleto ante esta nueva tecnología. La organización y el funcionamiento de un microprocesador no son perceptibles por los sentidos. Todo se reduce a una pequeña caja que contiene la función técnica; por lo tanto, el diseñador posee mucha más libertad para la configuración final del objeto.

Ante esta libertad formal el diseñador debería, con ayuda de la multidisciplinaridad actual del trabajo, humanizar más la técnica y cambiar la relación hombre-objeto.

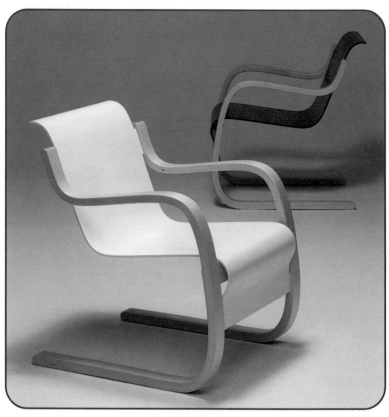

Armchair 42 *by Artek*

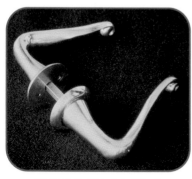

Manubrio Batlló *by Antoni Gaudí*

Armchair 41 Paimo *by Artek*

CONSTRUCTING OBJECTS

The history of design has been closely linked to architecture, but they have not evolved in tandem, as it was architecture that created the ground rules for industrial design, particularly in the case of furniture. At the beginning of the 20th century, with developments in architecture such as Modernism and, years later, the so-called Modern Movement, this relationship of dependency became increasingly evident.

The pioneering architects of the avant-garde formulated new concepts of space that made the furniture of the time obsolete. So, they decided to create for themselves furniture that was more in keeping with their innovative ideas. In this way, whether they intended to or not, they produced some of the 20th century's most important designs.

The outlines of this relationship between design and architecture changed radically in the 1980s with the design "boom" in the media. The major design companies hired prestigious architects to create their products, but these moves were entirely motivated by considerations of marketing.

DIE ENTWICKLUNG VON OBJEKTEN

Die Architektur und die Geschichte des Designs sind eng miteinander verbunden; es hat hier keine parallele Entwicklung stattgefunden, sondern die Architektur hat im Gegenteil die Richtlinien für das industrielle Design, und ganz besonders für das Mobiliar, geschaffen. Besonders deutlich wird diese Abhängigkeit zu Beginn des 20. Jahrhunderts im Modernismus und einige Jahre später in der sogenannten Modernen Bewegung der Architektur.

Die avantgardistischen Architekten schaffen neue Raumvorstellungen, die in deutlichem Gegensatz zu dem veralteten Mobiliar der Epoche stehen. Sie selbst kreieren Möbel, die mit ihren neuerungsfreudigen Ideen besser harmonieren, und wurden auf diese Weise – gewollt oder ungewollt – zu den Schöpfern der markantesten Stücke des 20. Jahrhundert.

Das Wesen dieser Beziehung ändert sich radikal in den 80er Jahren, als es zu einem „Designboom" in den Kommunikationsmedien kommt. Die großen Firmen stellen namhafte Architekten ein, die ihre Objekte ausschließlich zu Marketingzwecken signieren.

Tea Trolley 900 *by Artek*

CONSTRUIRE DES OBJETS

L'architecture et l'histoire du design ont entretenu une relation intime, sans connaître pour autant une évolution en parallèle, l'architecture ayant marqué la mesure pour le design industriel et plus spécialement le mobilier. C'est au commencement du XXème siècle avec le Modernisme et, quelques années plus tard, avec le Mouvement Moderne en architecture que cette relation de dépendance se fait patente.

Les architectes les plus avant-gardistes créent de nouvelles conceptions de l'espace qui contrastent avec le mobilier de l'époque, devenu obsolète. Ainsi, ils décident de concevoir eux-mêmes un mobilier plus en harmonie avec leurs propositions novatrices. De ce fait, intentionnellement ou non, ils sont responsables des pièces les plus significatives du XXème siècle.

Le schéma de cette relation change radicalement dans la décennie des années 80 avec le « boom » du design dans les médias. Les grandes entreprises embauchent des architectes de renom afin de signer leurs objets dans le cadre d'opérations de pur marketing.

High Kitchen Chair K56 *by Alvar Aalto. Artek*

Kitchen Chair 65
by Alvar Aalto. Artek

CONSTRUYENDO OBJETOS

La arquitectura y la historia del diseño han estado íntimamente relacionadas. No ha sido una evolución en paralelo, sino que es la arquitectura quien ha marcado las pautas del diseño industrial y en especial las del mobiliario. A principios del siglo XX con el Modernismo y años más tarde con el Movimiento Moderno en la arquitectura es cuando esta relación de dependencia se hace más patente.

Los arquitectos más vanguardistas crean nuevas concepciones del espacio que contrastan con el mobiliario obsoleto de la época. De esta manera deciden proyectar ellos mismos un mobiliario más acorde con sus innovadoras propuestas. Así, intencionadamente o no, se convierten en responsables de las piezas más significativas del siglo XX. El dibujo de esta relación cambia radicalmente en la década de los ochenta con el *boom* del diseño en los medios de comunicación. Las grandes empresas fichan a arquitectos de renombre para que firmen sus objetos en una operación puramente de marketing.

Packaging

Verpackung

Emballage

Packaging

Packaging

This is an example of the multidisciplinary work involved in the conception of a product. Here the industrial designer has to work alongside experts in graphic design to achieve a product that stands out, in the terms required, in an extensive market.

Verpackung

Dies ist ein Beispiel für den vielfältigen und verschiedene Disziplinen umfassenden Arbeitsprozess zur Konzeption eines Produktes. Hier muss der Industriedesigner Hand in Hand mit dem Grafikdesigner und dem Marketing arbeiten, um ein Produkt zu entwickeln, das sich unter bestimmten Voraussetzungen innerhalb eines breiten Marktes differenziert.

Emballage

C'est un exemple de travail pluridisciplinaire réalisé pour la conception d'un produit. Ici, le designer industriel doit travailler main dans la main avec le design graphique et le marketing afin d'obtenir un produit différencié, selon les termes qui nous importent, dans le cadre d'un vaste marché.

Packaging

Es un ejemplo del trabajo multidisciplinar que se realiza para la concepción de un producto. Aquí el diseñador industrial ha de trabajar de la mano del diseño gráfico y del marketing para poder lograr un producto que se diferencie, en los términos que nos interese, dentro de un amplio mercado.

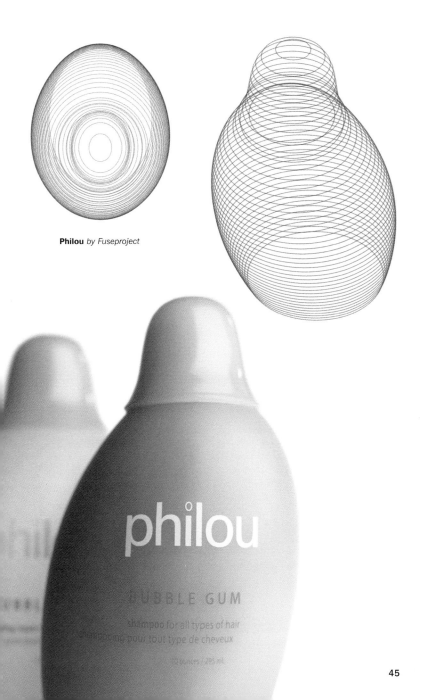

Philou by Fuseproject

45

Space Scent *by Fuseproject*

space scent 001

Body milk & shower gel containers "Le Feu d'Issey Light" *for Issey Miyake*
by Radi Designers

Perfume 09
by Fuseproject

Aquadove water bottle
by Emilio Ambasz + Associates Inc

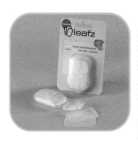

Compactcase for soap
by Rand Id

Stack CD Holder
for TDK by Ideo Europe

Oxo Grind It *by Oxo*

Oxo Grind It packaging
by Smart Design

Chocolate box for Nestlé

Twist'n Go cup *for Pepsi by Ideo Europe*

Range of measure
for Ricard
by Radi Designers

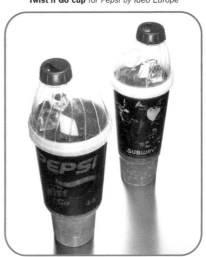

Bread knife *by Franz Güde.*
Franz Güde GmbH

Pharem *by Rand Id*

space scent 001

Improved convenient battery packages
by Noriko Himeda

The Design Process

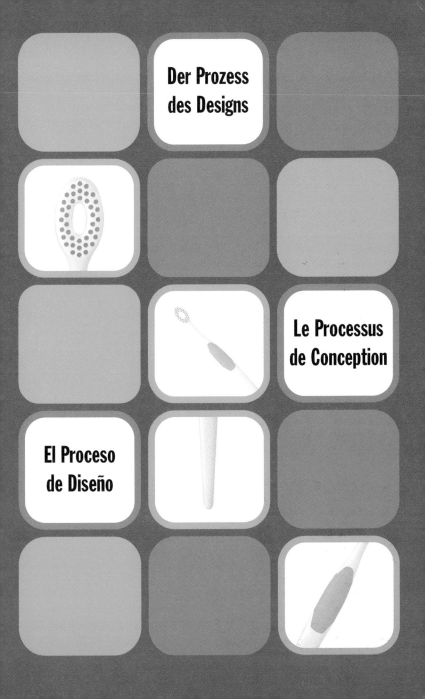

**Der Prozess
des Designs**

**Le Processus
de Conception**

**El Proceso
de Diseño**

The Design Process

Designing a new object involves a process of trial and error, starting with sketches, continuing with models and finishing with tests on prototypes. It is a long and laborious process, where inspiration serves for little without solid multidisciplinary groundwork.

Der Prozess des Designs

Das Design eines neuen Objektes erfordert eine Zeit der Versuche und der Irrtümer. Am Anfang stehen Zeichnungen, gefolgt von der Weiterentwicklung anhand eines Modells und den Versuchen an einem Prototyp. Es ist ein langer und arbeitsintensiver Prozess, bei dem die Inspiration nur Erfolg hat, wenn sie sich auf eine Arbeit vielfältiger Disziplinen stützen kann.

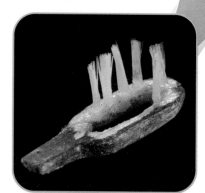

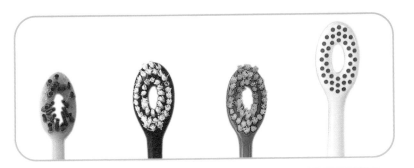

Le Processus de Conception
Concevoir un nouvel objet implique
un processus d'essais et d'erreurs.
Commencer par des dessins,
poursuivre avec des maquettes et
terminer par des essais de
prototypes. Le processus est long
et astreignant et l'inspiration est
de peu de recours si elle n'est pas
sous-tendue par un travail
pluridisciplinaire.

El Proceso del Diseño
Diseñar un nuevo objeto requiere un
proceso de ensayo y error.
Empezar con dibujos, continuar con
maquetas y acabar ensayando con
prototipos.
Es un proceso largo y trabajoso,
donde la inspiración sirve
de poco si no hay un trabajo
multidisciplinar detrás.

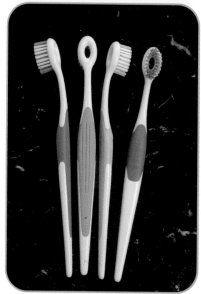

Ozone toothbrush *by Ozone Dental Products*

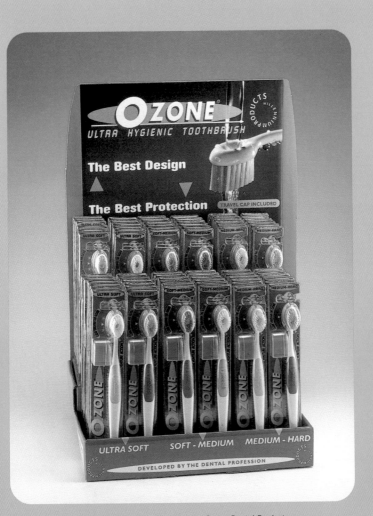

Ozone toothbrush packaging *by Ozone Dental Products*

Current Trends

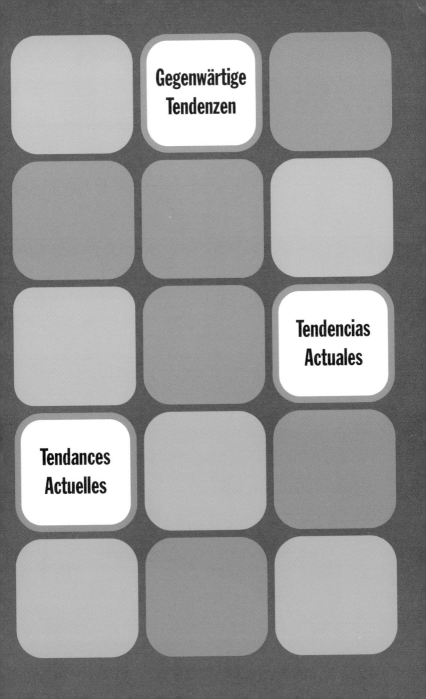

Gegenwärtige
Tendenzen

Tendencias
Actuales

Tendances
Actuelles

Current Trends

The form of objects does not only obey functional criteria. A well-designed object is capable of communicating something over and above its function—it has to be able to convey its intentions and establish a different relationship between subject and object. This is why we prefer some objects over others, even though they may serve the same purpose. The world of industrial design is also swayed by fashion; companies know this and launch products to follow the latest trends.

Gegenwärtige Tendenzen

Die Form der Objekte richtet sich nicht nur nach funktionellen Kriterien. Ein gut gestaltetes Objekt teilt etwas mit, das über seine Funktion hinausgeht. Es muss in der Lage sein, seine Absicht zu vermitteln, eine neue Beziehung zwischen Subjekt und Objekt herzustellen. So können wir uns vor verschiedenen Produkten gleicher Funktion bestimmten davon zuwenden und andere gar nicht beachten. Auch die Welt des industriellen Designs unterliegt der Mode; die Firmen sind sich dieser Tatsache bewusst und halten sich mit ihren Produkten an diese Tendenzen.

Leni by Büro für Form

Fluent chair series by Nick Jinkinson and Nick Paterson / Studio Platform

62

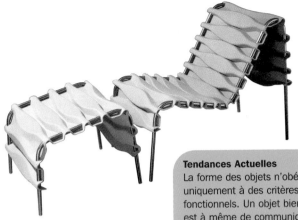

Tendances Actuelles

La forme des objets n'obéit pas uniquement à des critères fonctionnels. Un objet bien conçu est à même de communiquer autre chose que sa fonction : il doit être capable de communiquer son intention, d'établir une relation sujet-objet différente. Ainsi, devant plusieurs objets offrant les mêmes prestations, nous pouvons tomber amoureux de certains et pas d'autres. Le monde du design industriel est également sujet des modes, les entreprises en étant conscientes lancent leurs produits en suivant ces tendances.

Tendencias Actuales

La forma de los objetos no solamente obedece a criterios funcionales. Un objeto bien diseñado es capaz de comunicar algo más que su función, ha de ser capaz de comunicar su intención, de establecer una relación diferente entre sujeto y objeto. Así, ante diferentes productos que ofrecen las mismas prestaciones somos capaces de enamorarnos de unos y no de otros. El mundo del diseño industrial también está sujeto a las modas;las empresas lo saben y lanzan sus productos siguiendo estas tendencias.

63

Neo-retro

These are designs that combine the latest technology with a look inspired by American designs of the 1940s and 1950s, with their rounded, aerodynamic finishing. They emanate robustness and seem to promise that the products in question work exactly as they used to back in the good old days.

Neo-Retro

Diese Designs verbinden die modernsten Technologien mit dem Look des amerikanischen Designs in den vierziger und den fünfziger Jahren, das durch sein abgerundetes und aerodynamisches Styling gekennzeichnet ist. Die Produkte machen einen robusten Eindruck und vermitteln das Gefühl der Zuverlässigkeit in ihrer Funktionsweise wie in der „guten alten Zeit."

Oby Light I & II *by Büro für Form*

Copacabana *by Eero Arnio*

Eat & Lounge
by Büro für Form

Colema by Marc Newson

Néo-rétro

Il s'agit des designs qui allient les technologies de dernière génération à un look inspiré des créations américaines de la fin des années quarante et des années cinquante. C'est le Styling, avec ses finitions arrondies et aérodynamiques. Elles ont un air robuste mais semblent garantir que le produit fonctionne comme ceux d'avant.

Neo-retro

Son diseños que mezclan las últimas tecnologías con una apariencia inspirada en diseños americanos de finales de los cuarenta y los cincuenta, como el Styling, con sus acabados redondeados y aerodinámicos. Poseen un cierto aire robusto, pero parecen garantizar que es un producto que funciona como los de antes.

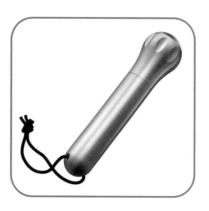

Floss Torch *by Marc Newson*

Hi-tech

Here the product design is very clean and limpid, bordering on minimalism. It gives an image of precision, with no superfluous or missing details and the most appropriate materials.

High-tech

Dieses Produktdesign ist sauber, klar, fast schon minimalistisch. Es vermittelt den Eindruck von Präzision und kein Detail ist zu viel oder zu wenig. Für seine Umsetzung verwendet man nur die am besten geeigneten Materialien.

The pillow alarm clock uses an unusual principal to wake you up —that of natural light. A gentle increase in intensity of light wakes the user up refreshed to a simulated sunrise.

A nicer way to wake up
by Studio Platform

High-tech
Ici le design du produit est très soigné, clair, flirtant avec le minimalisme. Il offre une image de précision, où aucun détail n'est de trop ou ne manque, et ses matériaux ont été choisis pour leur adéquation.

High-tech
Aquí el diseño del producto es muy limpio, claro, rozando el minimalismo. Da una imagen de precisión, donde no le sobra ni le falta ningún detalle y sus materiales han sido escogidos por su idoneidad.

Pod Watch *by Marc Newson*

Organic
This is an approach to design inspired by the shapes found in nature, based on sinuous curves that soften forms. This type of design requires a mastery of materials to be able to achieve the desired effects.

Organisches Design
Hier wird das Design durch die Formen der Natur inspiriert und basiert auf sinusförmigen Linien, die die Formen sanfter gestalten. Dieses Design verlangt profunde Kenntnisse des Materialverhaltens, um die gewünschte Wirkung zu erreichen.

Organique
C'est une forme de design inspirée des formes de la nature. Elle repose sur la sinuosité des courbes afin d'adoucir les formes. Ce type de conception nécessite une ample connaissance du comportement des matériaux pour obtenir l'effet désiré.

Orgánico
Es un enfoque del diseño inspirado en las formas de la naturaleza. Se basa en la sinuosidad de las curvas para suavizar las formas. Es un diseño que requiere un gran conocimiento del comportamiento de los materiales para conseguir el efecto deseado.

Ashtray by Marc Newson

Desk Sail
by Studio Platform

Mystery Clock *by Marc Newson*

Prog Oggetto Vase *by Marc Newson*

Soft-plastics

Innovations in technopolymers have made it possible to create objects with the most surprising and amusing forms. This trend no longer embraces only small accessories but has also reached the world of technological equipment to provide an image of accessibility and convenience.

Weichplastik

Durch die Neuerungen auf dem Gebiet der technischen Polymere konnten Objekte in den überraschendsten und witzigsten Formen hergestellt werden. Diese Tendenz konzentriert sich nicht nur auf kleine Zubehörteile, sondern hat auch in die Welt der technologischen Geräte Eingang gefunden, wo sie die Vorstellung von Zugänglichkeit und leichter Bedienung vermittelt.

Dish Doctor *by Magis Design*

Plastiques Mous

L'innovation dans les techno-polymères a permis de créer des objets aux formes les plus surprenantes et amusantes. Ce courant ne se centre pas uniquement sur le design des petits accessoires mais a pu s'insérer également dans le monde des appareils plus technologiques afin de leur conférer une image d'accessibilité et de convivialité.

Soft-plastics

La innovación en los tecnopolímeros ha permitido crear objetos de las formas más sorprendentes y divertidas. Esta corriente no sólo se centra en el diseño de pequeños accesorios sino que también ha llegado al mundo de los aparatos más tecnológicos para dar una imagen de accesibilidad y fácil manejo.

Wearable Hanger by Studio Platform

Chaise Volante
by Büro für Form

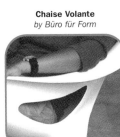

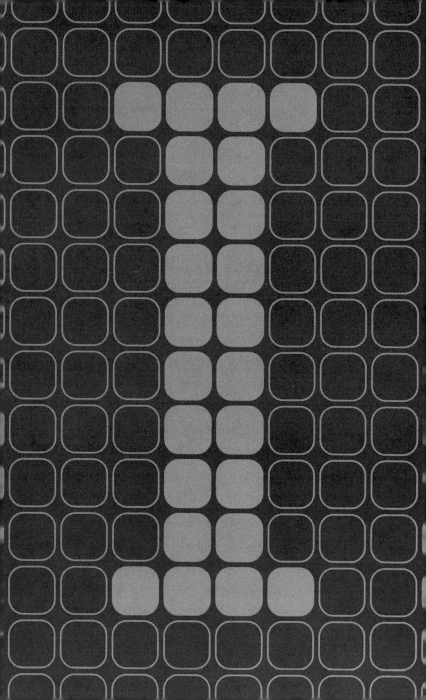

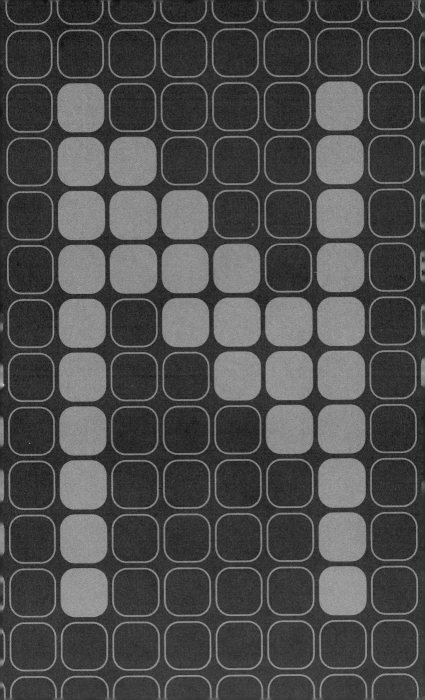

Job

Bureau

Office

Büro

Oficina

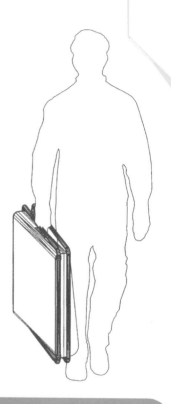

Take Away by *Beat Karrer*

Office

The development of the new technologies has brought changes in the way we work. Values such as mobility and flexibility have come into play to break free from the restrictions of the traditional office and enhance personal creativity and productivity.

Büro

Die Entwicklung neuer Technologien brachte eine Veränderung der Arbeit und ihrer Durchführung mit sich. Um sich von den traditionellen Konzepten des Büros zu lösen, wandte man sich neuen Werten wie Beweglichkeit und Flexibilität zu, die die Kreativität und Produktivität des Einzelnen steigern.

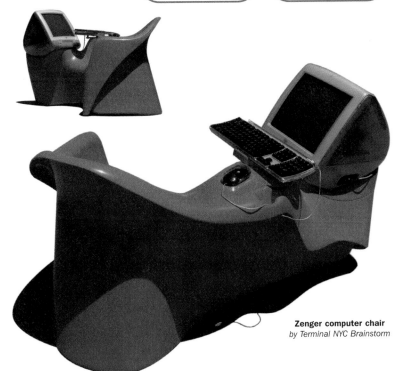

Zenger computer chair
by Terminal NYC Brainstorm

Scooter *by Beat Karrer*

The Fingertip Table
by Thom Faulders.
Beige Architecture Design

Bureau

L'évolution des nouvelles technologies a permis que le travail et la forme de le réaliser évoluent. Pour rompre le lien avec les anciens bureaux, des valeurs ont été adoptées ainsi la mobilité et la flexibilité qui permettent d'impulser la créativité et la productivité d'un individu.

Oficina

La evolución de las nuevas tecnologías ha permitido que el trabajo y la forma de realizarlo estén cambiando. Para romper con la atadura de la antigua oficina se adoptan valores como la movilidad y la flexibilidad, que permiten potenciar la creatividad y productividad del individuo.

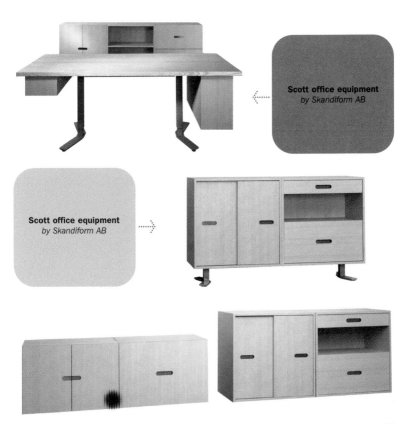

Scott office equipment
by Skandiform AB

Scott office equipment
by Skandiform AB

Transit by Skandiform AB

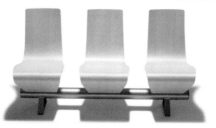

Kart chair *for Vecta by Ideo Europe*

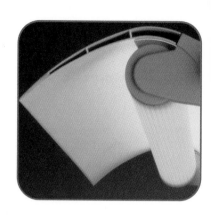

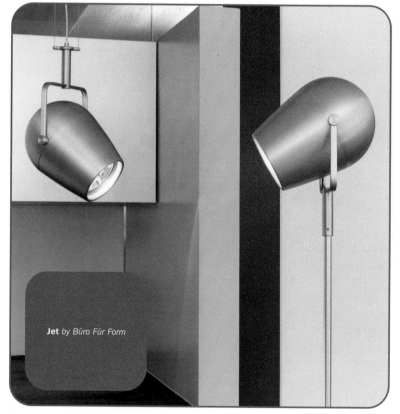

Jet *by Büro Für Form*

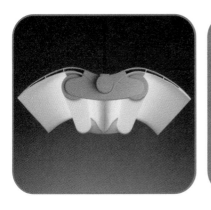

Expanded
by King & Miranda Design

Alba *by King & Miranda Design*

Furniture
Office furniture not only aims for flexibility to adapt to any need that might arise; it has also become an important element in corporate identity, and even in company strategy.

Mobiliar
Neben der Suche nach Flexibilität, je nach den auftretenden Bedürfnissen, wurde das Mobiliar auch zu einem wichtigen Element der Identität einer Firma und ihrer Strategie.

Mobilier
Outre le fait de rechercher la flexibilité, selon l'apparition des besoins, le mobilier s'est converti en un élément important de l'identité de l'entreprise ainsi que de sa stratégie.

Mobiliario
Además de buscar la flexibilidad según vayan apareciendo las necesidades, el mobiliario se ha convertido en un elemento importante en la identidad corporativa así como en la estrategia de la empresa.

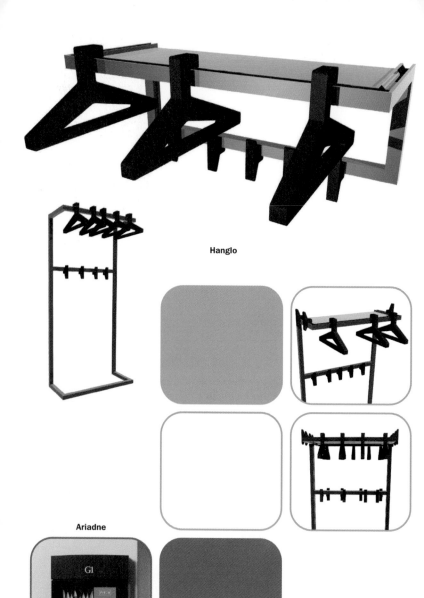

Hanglo

Ariadne

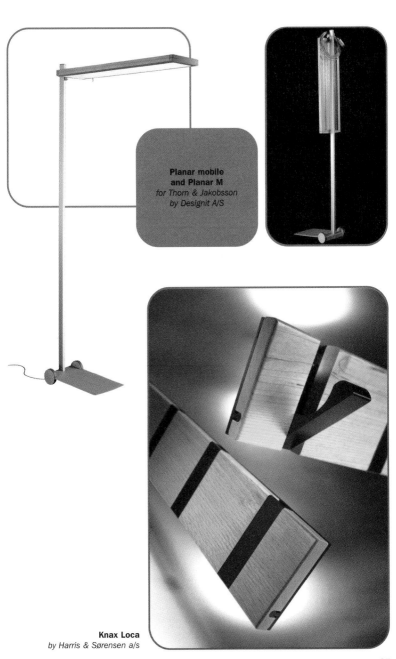

**Planar mobile
and Planar M**
*for Thorn & Jakobsson
by Designit A/S*

Knax Loca
by Harris & Sørensen a/s

85

Magazine Rack
by Eyecatcher

Design by Achilles Associates bvba

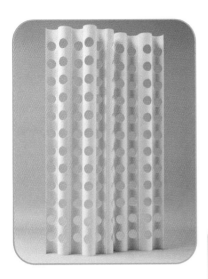

Perf Screen
by Thom Faulders.
Beige Architecture Design

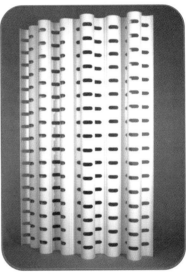

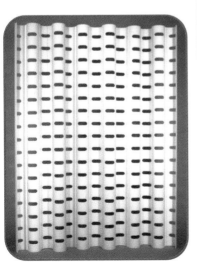

Boston Executive Stand-up Stapler
for Hunt by Ecco

Boston Orca Mini Stand-up Stapler
for Hunt by Ecco

Office Materials
The wide range of products currently on offer turn their back on the impersonal designs of the past in favour of the creation of personalized work spaces.

Büromaterial
Heutzutage erlaubt die große Produktpalette die persönliche Gestaltung des Arbeitsraumes und die Abkehr von unpersönlichen Designs früherer Zeiten.

Matériel de Bureau
Fuyant les anciens designs impersonnels, la vaste gamme de produits permet actuellement de personnaliser l'espace de travail.

Material de Oficina
Huyendo de los antiguos diseños impersonales, actualmente una amplia gama de productos permite personalizar el espacio de trabajo.

Design by Lexon

Design by Lexon

Gonzales by Koziol

Speedy by Koziol

90

Curly *by Koziol*

Remarkable Recycled Pencil
by Noel Murphy. Remarkable

Soft Edge Ruler *for Back Crubtt by Ecco*

Ashtray *by Eyecatcher*

Creates design to people for whom aesthetics and simplicity is a way of life.

Yott lighted lens *by Frédéric Lintz. Lexon*

Design by Lexon

Design by Lexon

Design by Lexon

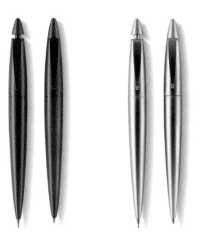

Design by Lexon

Design by Lexon

Design by Lexon

Atlantic *for Modulex by Designit A/S*

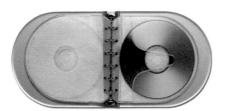

Design by Lexon

97

Plobb! *by Beat Karrer*

Lounge waste paper basket *by Almén:gest Design*

Samsonite-Focus attachecase
by Enthoven Associates

FedEx courier tools
by Ziba

The FedEx courier tools reflect the principles of a functional, flexible and user-friendly design language.

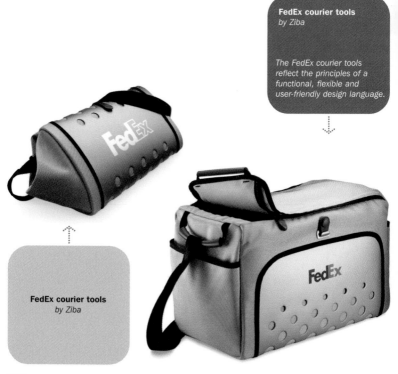

FedEx courier tools
by Ziba

FedEx courier tools *by Ziba*

FedEx courier tools *by Ziba*

Telecommunications

The convenience and low costs of the new information and communication technologies enhance personal creativity as the more mechanical tasks can be performed by computers.

Fernmeldewesen

Durch die niedrigen Kosten und die einfache Bedienung der neuen Informations- und Kommunikationstechnologien ist es möglich, die Kreativität des Einzelnen zu potenzieren, da sogar die einfachsten mechanischen Prozesse mit dem Computer gelöst werden können.

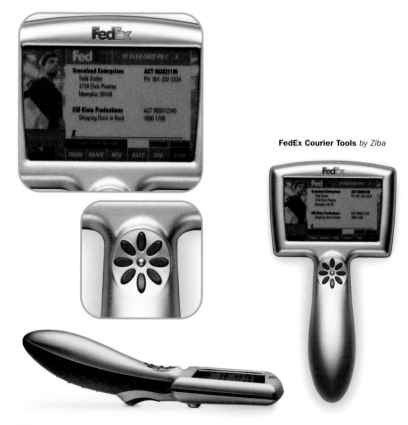

FedEx Courier Tools *by Ziba*

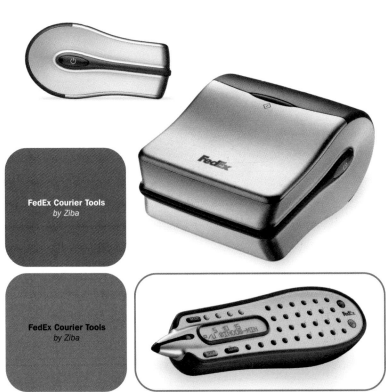

FedEx Courier Tools
by Ziba

FedEx Courier Tools
by Ziba

Télécommunications
Le faible coût et l'utilisation simplifiée des nouvelles technologies de l'information et de communication permettent de susciter la créativité personnelle, les processus les plus mécaniques pouvant être résolus par les ordinateurs.

Telecomunicaciones
El bajo coste y fácil manejo de las nuevas tecnologías de la información y comunicación permiten potenciar la creatividad de la persona ya que los procesos más mecánicos pueden ser solucionados por computadoras.

Beocom3
*for Bang & Olufsen A/S
by Designit A/S*

Topco
by Achilles Associates bvba

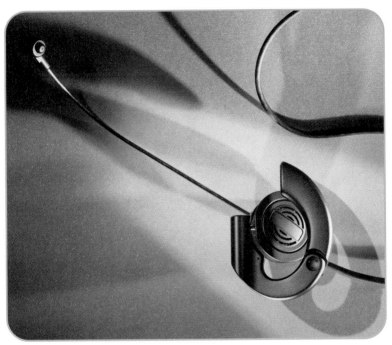

Headset *for Unex by Ideo Europe*

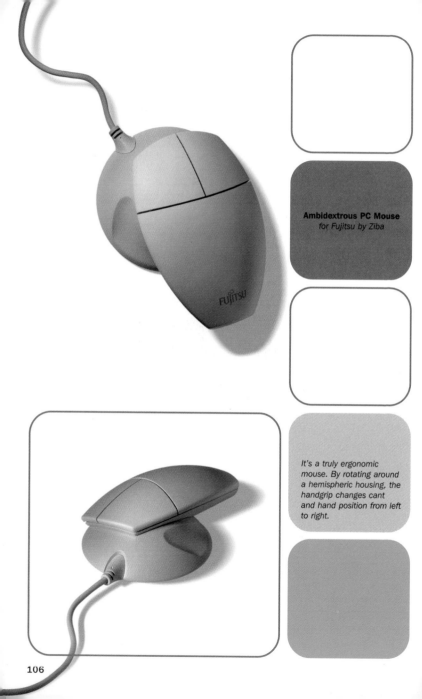

Ambidextrous PC Mouse
for Fujitsu by Ziba

It's a truly ergonomic mouse. By rotating around a hemispheric housing, the handgrip changes cant and hand position from left to right.

HP Pavillion 2000
by Fuseproject

New Concept PC
for Intel by Ziba

New Concept PC ·······›
for Intel by Ziba

*These PC CPUs are truely
plug-and-play products.
It's a limited edition with
the goal to stimulate interest
among PC manufacturers
and the media for a new,
simple design.*

108

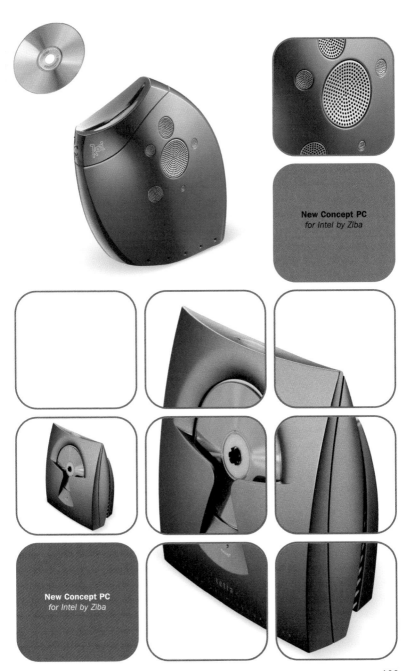

New Concept PC
for Intel by Ziba

New Concept PC
for Intel by Ziba

109

LX Print Server
for EFI by Ziba

E-Scanner *by Smart Design*

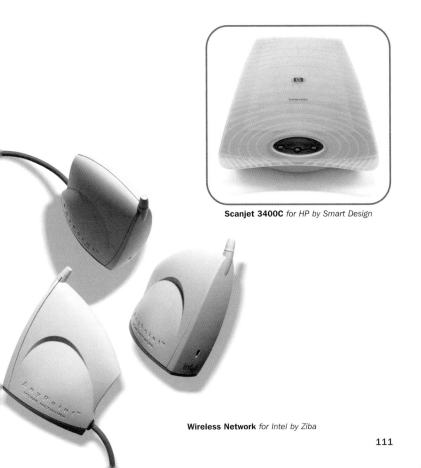

Scanjet 3400C *for HP by Smart Design*

Wireless Network *for Intel by Ziba*

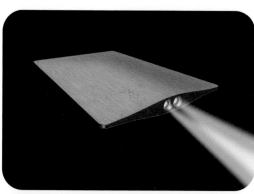

Econ *by Iain Sinclair*

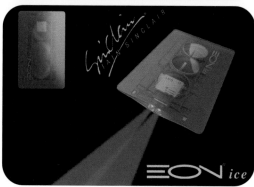

Design by Lexon

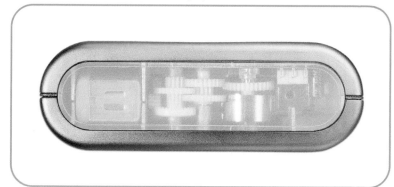

Photochemical Recycling System
for PRS by Ideo Europe

Printer Brivo
by Ideo Europe

PC Speakers *for Kenwood by Ziba*

Sound Chamber *for Pioneer by Ziba*

Small Touch
by Designit A/S

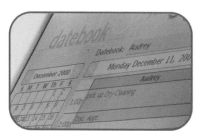

Ergo Audrey datebook
by Ideo Europe

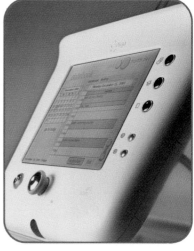

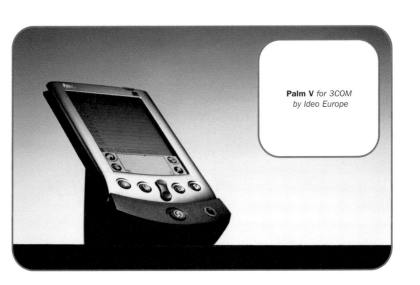

Palm V *for 3COM*
by Ideo Europe

Treo *for Handspring*
by Ideo Europe

Job

Herramientas

Tools

Werkzeuge

Outils

Design by Lexon

Design by Lexon

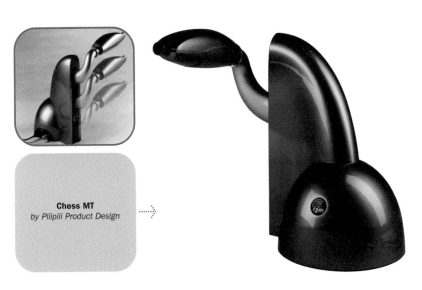

Chess MT
by Pilipili Product Design

Design by Lexon

Design by Lexon

121

Biojector non-invasive injection system
for Biotect by Ziba

Medical Materials

The function is of overriding importance in medicine dispensers and diagnostic equipment, but a good design not only makes them more convenient to use but also tones down their foreboding appearance.

Medizinisches Material

Sowohl bei den Geräten für die Verwaltung von Medikamenten als auch bei jenen der Diagnose ist die Funktion für die Gestaltung des Objekts maßgeblich. Ein gelungenes Design kann jedoch die Bedienung vereinfachen und das angsteinflößende Aussehen mildern.

Matériel Médical

Tant pour les équipements de gestion des médicaments comme pour les appareils de diagnostic, la forme est dictée par la fonction. Cependant, un bon design peut rendre plus pratique et adoucir un aspect menaçant.

Biojector *by Ziba*

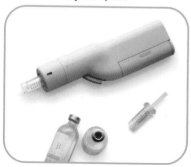

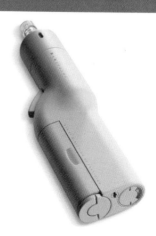

Nextep Contour Leg Walker
for Johnson & Johnson Orthopaedic
by Ecco

*It's a bracing device indicated
for the treatment of stable
ankle fracture and general
orthopaedic injuries. The
aesthetic appeal promotes it's
continued use through time.*

Material Médico

Tanto en aparatos de administración
de medicamentos como en equipos
de diagnóstico la función es la que
rige el objeto, pero un buen diseño
puede hacer más cómodo y suavizar
su aspecto amenazador.

Vistalab Pipette *by Frog Design*

123

Fedesa by Design (Pool)

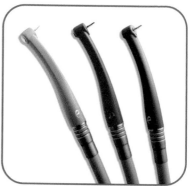

Dental Drill for Oral by Ziba

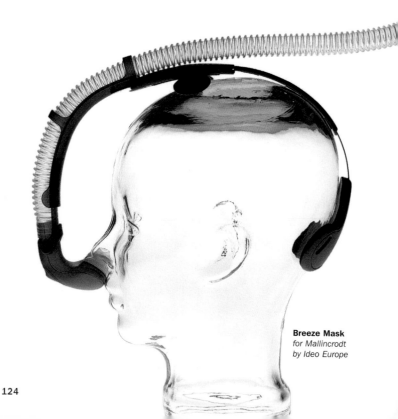

Breeze Mask
for Mallincrodt
by Ideo Europe

Design by Lexon

Design by Lexon

125

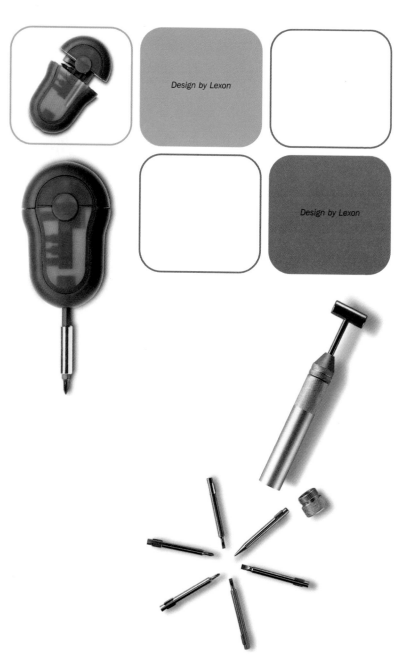

Design by Lexon

Design by Lexon

Tools

If one of the aims of industrial design is to solve problems, this function can be seen most clearly in the case of tools. Their design is intended to make them easier and more convenient to use.

Werkzeuge

Wenn eine der Zielsetzungen des industriellen Designs darin besteht, Probleme zu lösen oder Produkte zu verbessern, so ist dies auf dem Sektor der Werkzeuge am deutlichsten zu sehen. Diese Designs sind dazu gedacht, die Arbeit für die Anwender leichter und angenehmer zu gestalten.

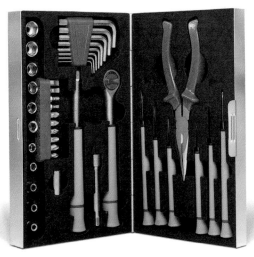

Design by Lexon

Outils

Si l'un des objectifs du design industriel est de résoudre les problèmes, voire d'apporter une amélioration, c'est bien dans le domaine des outils que cette fonction se fait le plus patente. Il s'agit de designs pensés pour faciliter et rendre plus agréable le travail pour ses acteurs.

Herramientas

Si uno de los objetivos del diseño industrial es el de solucionar o mejorar problemas, es en el campo de las herramientas donde esta función se puede ver más claramente. Son diseños pensados para hacer más fácil y agradable el trabajo a sus usuarios.

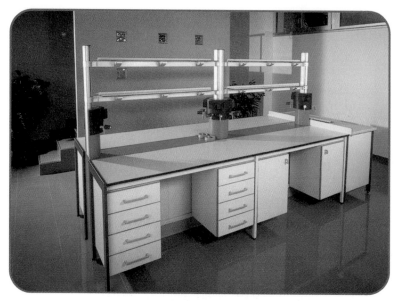

Burdinola *by Tres D*

NovoRapid Flexpen
by Designit A/S

Design by Oxo

Allround Brushrest *by Pilipili Product Design*

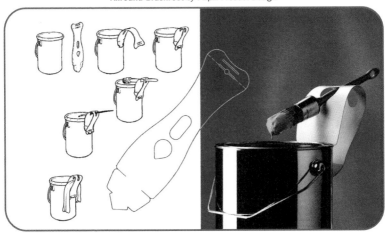

129

Machinery

This is a good example of multidisciplinary creation, in which engineering and design have to work together to achieve be combined reliable and efficient machines.

Maschinen

Dies ist ein Beispiel für das Zusammenwirken von Disziplinen – Planung und Design müssen hier Hand in Hand gehen, um vertrauenswürdige und leistungsstarke Maschinen herzustellen.

John Deere Garden Tractor
by Henry Dreyfuss Associates

John Deere GT 225
by Henry Dreyfuss Associates

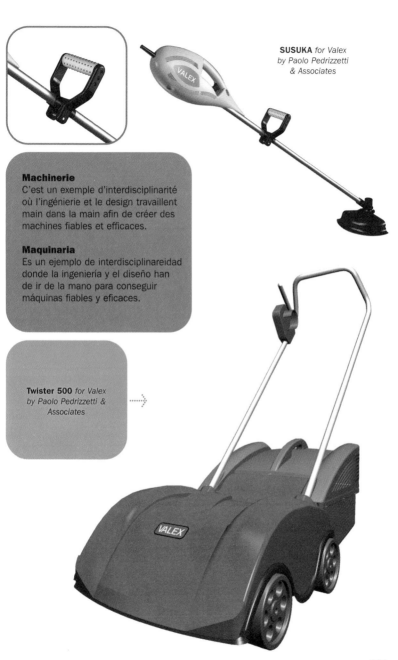

SUSUKA *for Valex*
by Paolo Pedrizzetti
& Associates

Machinerie
C'est un exemple d'interdisciplinarité où l'ingénierie et le design travaillent main dans la main afin de créer des machines fiables et efficaces.

Maquinaria
Es un ejemplo de interdisciplinareidad donde la ingeniería y el diseño han de ir de la mano para conseguir máquinas fiables y eficaces.

Twister 500 *for Valex*
by Paolo Pedrizzetti &
Associates

131

Hyster lift truck operator's station *by Henry Dreyfuss Associates*

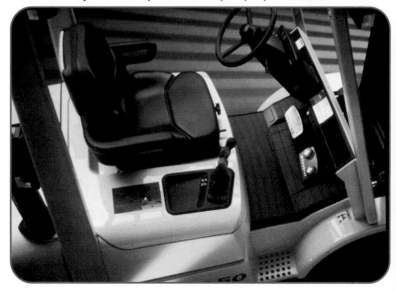

132

Iberdrola Recarga Battery
by Tres D

Adher Electrónica
by Tres D

Design by Tres D

Mebusa Control Nº
by Tres D

Utensils

The esthetic virtues of an object are often dictated by its functionality and convenience. These demand exhaustive anthropometric and ergonomic studies to make sure that objects are safe and easy to use, as well as being beautiful.

Handwerkszeuge

Oft hängt die Schönheit eines Gegenstandes von seiner Funktionalität und seiner bequemen Handhabung ab. Die Herstellung schöner Objekte, die leicht und sicher zu benutzen sind, erfordert gründliche Untersuchungen der Anthropometrie und der Ergonomie.

Ustensiles

Souvent, la beauté d'un objet est dictée par sa fonctionnalité et sa commodité d'usage. S'imposent des études anthropométriques et ergonomiques exhaustives afin d'obtenir des objets simples à utiliser, sûrs et beaux.

Utensilios

Muchas veces la belleza de un objeto viene dictada por su funcionalidad y comodidad de uso. Esto requiere unos exhaustivos estudios de antropometría y ergonomía para conseguir objetos de uso fácil, seguros y bellos.

SCS wallet *by Mormedi*

SCS TPV *by Mormedi*

Tesa *by Tres D*

Goitek by Tres D

BodyGem by Ideo Europe

Design by Design (Pool)

InDuo by Designit A/S

135

Logitech Kidz Mouse *by Frog Design*

Design by Lexon

Televic Axio *for Televic by Pilipili Product Design*

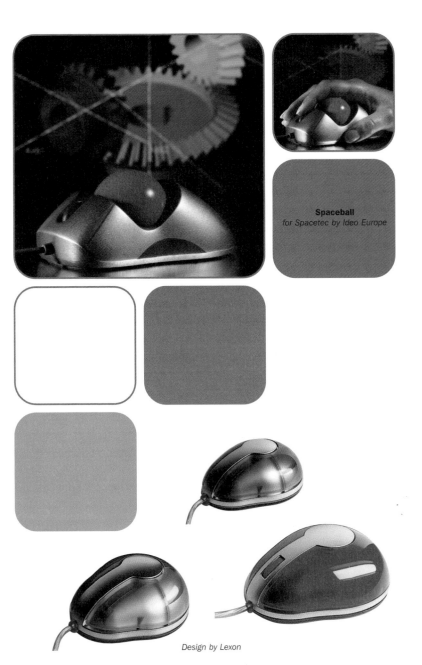

Spaceball
for Spacetec by Ideo Europe

Design by Lexon

137

Vadem Cliolaptop computer
by Frog Design

This laptop is winner of 1999 silver IDEA award in the business and industrial category.

Keyboard *by Apple*

Power Macintosh G4
by Apple

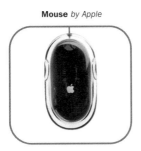

Ipod *by Apple*

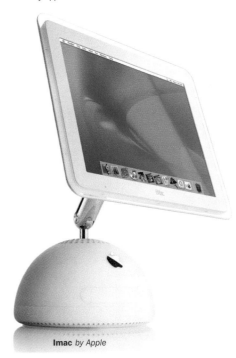

Imac *by Apple*

Mouse *by Apple*

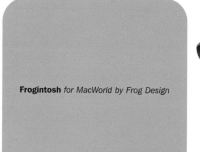

Frogintosh for *MacWorld by Frog Design*

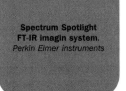

**Spectrum Spotlight
FT-IR imagin system.**
Perkin Elmer instruments

Design by Ideo Europe

◁······ **O2 ISIS**
by Achilles Associtates bvba

Home

Elektrogeräte

Pequeños Electrodomésticos

Electrical Appliances

Electroménagers

Cuisinart electrics *by Smart Design*

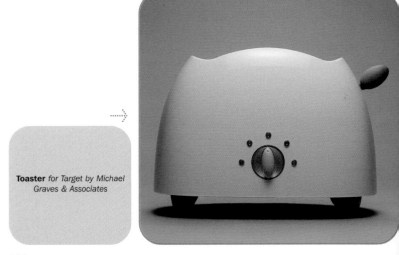

Toaster *for Target by Michael Graves & Associates*

Electrical Appliances
In the 20th century this field spawned countless inventions and developments. These products are jointly created by engineers, designers and marketing experts to make them as attractive as possible in every way.

Elektrogeräte
Während des vergangenen Jahrhunderts war die Welt der Elektrogeräte ein fruchtbarer Boden für Erfindungen und spätere Verbesserungen. Die Produkte wurden in enger Zusammenarbeit zwischen Ingenieuren, Designern und Marketing-Spezialisten entwickelt, mit dem Ziel, sie so attraktiv wie möglich zu gestalten.

Électroménagers
Durant le siècle passé, le monde de l'électroménager a été le champ de prédilection des inventions et des innovations postérieures. Les produits viennent developés conjointement par les ingénieurs, les designers et les experts en marketing afin d'obtenir des objets aussi séduisants que possibles.

Electrodomésticos
Durante el siglo pasado el mundo de los electrodomésticos ha sido campo abonado para los inventos y sus posteriores innovaciones. Los productos se desarrollan conjuntamente entre ingenieros, diseñadores y expertos en marketing para lograr objetos cuanto más atractivos mejor.

Titan washing machine
by TKO Design

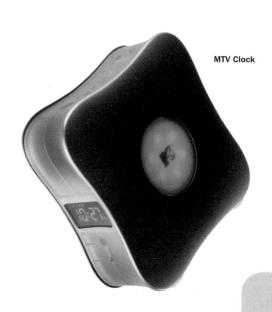

MTV Clock

SpinaLogic 1000

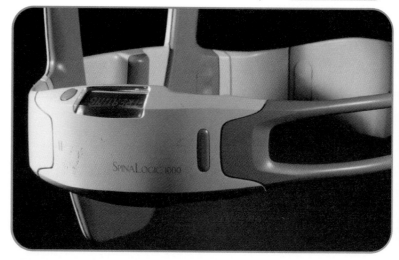

Orangex Juicer
by Smart Design

"A persons' emotional connection to a product comes as much from its function as it does from its form." (Smart Design)

Domestic Chores
Smart products are designed to help families in their monotonous everyday tasks.

Häusliche Aufgaben
Intelligente Produkte werden entwickelt, um ihren Benutzern die Monotonie ihrer alltäglichen Arbeiten zu nehmen.

Tâches Ménagères
Des produits intelligents sont conçus pour aider les personnes et les familles dans les tâches quotidiennes si monotones.

Tareas Domésticas
Se diseñan productos inteligentes que ayuden a las personas y a las familias en las monótonas tareas diarias.

BBC Digital Radio
by Ideo Europe

National Panasonic
by Design (Pool)

National Panasonic
by Design (Pool)

Q Concept *for Hoover by Ideo Europe*

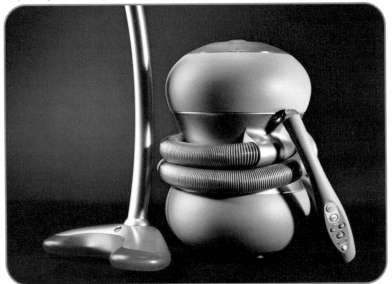

149

Design by Lexon

Design by Lexon

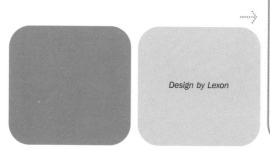

Design by Lexon

Design by Lexon

Design by Lexon

Entertainment

Today's products offer more and more possibilities. Designers are faced with the difficult task of making their technological functions clear and concise, while also making them visually attractive.

Unterhaltung

Die Leistungen der Produkte werden immer vielfältiger und besser. Es ist nicht leicht, das Gleichgewicht zwischen einer klaren und präzisen Technologie und der für das Produkt angestrebten Ästhetik herzustellen.

Divertissement

Les produits offrent des prestations à chaque fois meilleures et plus nombreuses. L'équilibre difficile porte sur la transmission de la technologie qu'ils contiennent, de manière claire et concise, alliée à l'esthétique auquel peut prétendre le produit.

Entretenimiento

Los productos tienen cada vez más y mejores prestaciones. El difícil equilibrio está en transmitir la tecnología que contienen de una manera clara y concisa, uniéndola a la estética que se pretende dar al producto.

THX home theater system *for Kenwood by Ziba*

Design by Lexon

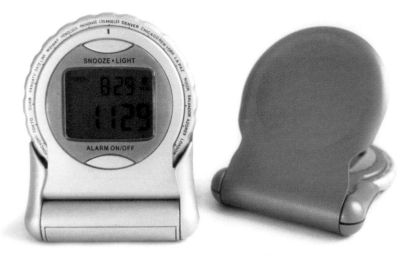

World Time clock *for One World Enterprises by Prime Studio Inc.*

Design by Lexon

Design by Lexon

Design by Lexon

Radio alarm clock
by Lexon

Design by Lexon

Design by Lexon

Design by Lexon

157

Soft Fan *by Priestman Goode*

Flowerpower
by Design Partners

Flowerpower works with new materials like soft but solid foam and gives an excellent and comfortable breeze in hot summer.

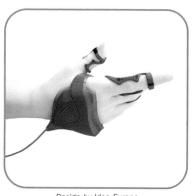

Design by Ideo Europe

Switch light-switch
by Radi Designers

Patrizia door-bell
by Radi Designers

DC 03 *by Dyson*

The lightest upright with the loss of suction is based on Dual Cyclone Technology.

DC 04 *by Dyson*

Highest performance and a slim profile to help clean under low furniture.

DC 04 *by Dyson*

Perfect combination of modern design and highest usability.

DC 05 *by Dyson*

161

Home

Möbel

Meubles

Muebles

Furniture

Carbon Lounge by *Fuseproject*

Furniture design is derived from four basic elements: the chair, the table, the chest and the bed. Today's furniture is passing through a phase marked by a diversity of influences.

Das Möbeldesign lässt sich auf vier Grundelemente zurückführen: Den Stuhl, den Tisch, die Truhe und das Bett. Gegenwärtig durchlebt die Welt der Möbel eine Phase, die von einer Vielzahl von Einflüssen gekennzeichet ist.

The thin and ultra-light carbon lounge chair fluidly negotiates continous shapes and carbon technology construction.

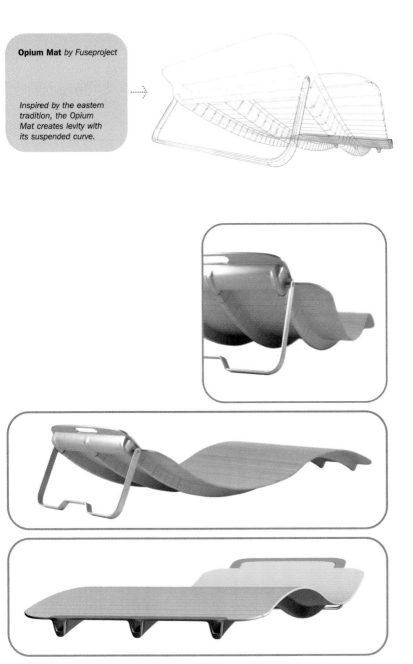

Opium Mat *by Fuseproject*

Inspired by the eastern tradition, the Opium Mat creates levity with its suspended curve.

165

Plastic-Bone
by Jean-Marie Massaud.
Magis Design

An example of a new way
to use plastic. This is a
chair made of plastic but
it is not a plastic chair!

Les meubles dérivent tous de
quatre prototypes originels: la chaise,
la table, le coffre et le lit.
Actuellement, le monde du meuble
est marqué par la disparité des
influences.

Puede afirmarse que todos los
muebles derivan de cuatro tipos
originarios: la silla, la mesa, el arca y
la cama. Actualmente el mundo del
mueble vive una etapa marcada por
la disparidad de influencias.

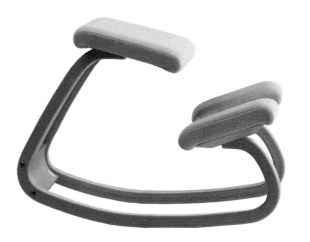

Variable *by Stokke*

Gravity *by Stokke*

SMD Chair
by Mac-interactive

169

Sussex by Robin Day.
Magis Design

A seating system, based on
a single plastic air-moulded
element, allowing the
construction of modular
seating units.

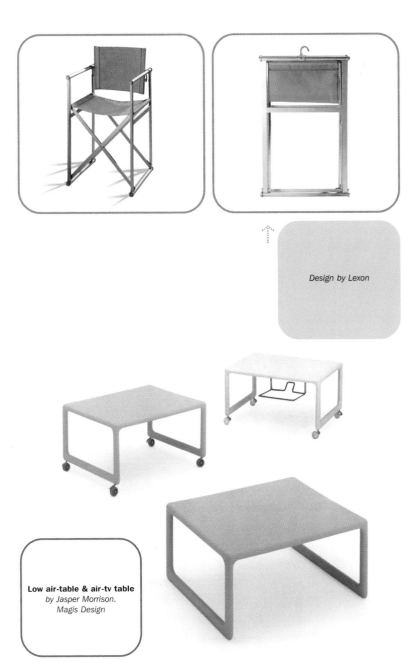

Design by Lexon

Low air-table & air-tv table
by Jasper Morrison.
Magis Design

171

It's only slightly larger than a chair. The half frame and the arm-rest are the same for both sides, meaning that only two moulds are needed to make the four parts of the chair. A perfect synthesis of aesthetic quality and industrial design technique.

Cinecitta *by Enzo Mari and Allessio Bozzer. Magis Design*

Folding air-chair
by Jasper Morrison.
Magis Design

A folding chair in
polyprophylene with glass
fibre addeds. Air moulded.

173

Nic Chair by Werner Aissinger. Magis Design

The biggest challange of this project was not so much in the seat but in the working out of the tubular steel structure. The result is a top-quality, comfortable chair.

Butterfly *by Karim Rashid.*
Magis Design

Almost everything is permissible, although the designer, despite of enjoying absolute freedom at the drawing board, is in reality often overruled by the norms established by the market, by practical considerations and by prevailing trends. Furniture design is undoubtly closely related to creative disciplines and art though.

Fast alles ist erlaubt und der Designer, der bei seiner kreativen Tätigkeit über absolute Freiheit verfügt, sieht sich oft an die Normen des Marktes, der Gesellschaft, der Ansprüche und der gerade herrschenden Moden gebunden. Das Möbeldesign ist zweifellos eng mit der Kunst verbunden.

Cependant, le designer, même s'il dispose d'une liberté absolue lorsqu'il crée, doit souvent se soumettre aux normes que le marché, la societé, les nécessités et les modes lui imposent. Le meuble quand même entretient une relation étroite avec les arts en général.

Casi todo está permitido, y el diseñador, a pesar de disponer de libertad absoluta de movimientos a la hora de crear, a menudo se encuentra supeditado a las normas que el mercado, la sociedad, las necesidades y las tendencias le marcan. El diseño de muebles sin duda se relaciona con disciplinas creativas y artísticas.

My 080 bed table
by Michael Young. Magis Design

**My 082 table &
My 083 nesting tables**
*by Michael Young.
Magis Design*

*Mixture of materials: the
frame is in polyprophylen,
connecting tubes in
chromed steel and the
top in tempered glases.*

Tam Tam low table & low stool *by Matteo Thun. Magis Design*

Made from polyethylene, this arrangement is suitable for outdoor use.

**Yogi Family low chair,
bench & low table**
*by Michael Young.
Magis Design*

One Family
by Konstantin Grcic.
Magis Design

The die-cast made products are simply extraordinary. The seat can also be mounted on a half-conical concrete base.

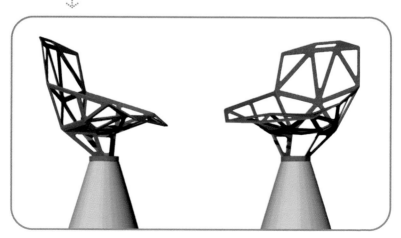

Nimrod *by Marc Newson. Magis Design*

Pankassa
*by Denis Santachiara.
Magis Design*

Extrem Lite *by Stokke*

Using plywood as the seat material, this chair shows a design, followed a straight aesthetic formula which brought up a splendid chair!

Lyra Low Chair
by Magis Design

Mariolino chair & side table *by Enzo Mari. Magis Design*

Stacking furniture that is perfect wherever space is a problem.

183

Alo tables *by Karim Rashid.*
Magis Design

Alo chair *by Karim Rashid.*
Magis Design

A chair in a graceful tone,
subdued and tranquil with a
fine and concrete detailing.

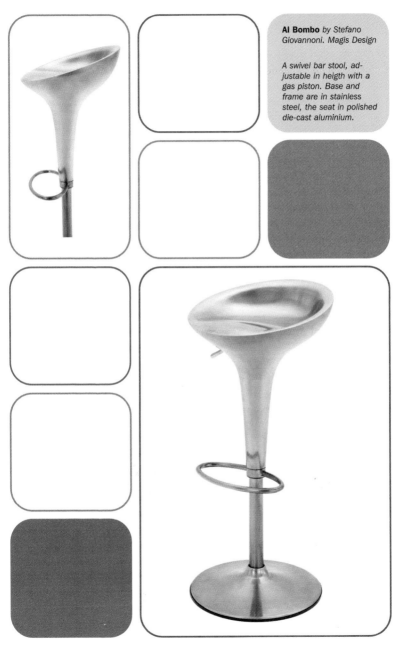

Al Bombo by *Stefano Giovannoni. Magis Design*

A swivel bar stool, adjustable in heigth with a gas piston. Base and frame are in stainless steel, the seat in polished die-cast aluminium.

185

Bombo chair & Big Bombo
*by Stefano Giovannoni.
Magis Design*

*Chairs' and tables' bases
are in chromed steal.
The chair is adjustable in
heigth and available in
eleven different colors.*

Bombo low chair & foot stool *by Stefano Giovannoni. Magis Design*

Base and seat material are in standard injection-moulded ABS. Cushion in polyethylene with leather or wool covering.

Bombo Table Ellittico *by Stefano Giovannoni. Magis Design*

Moontable by Fuseproject

Some designs are conceived as decorative elements as much as functional pieces of furniture and lie on the subtle borderline between sculpture and domestic objects. The visual appearance of these pieces is very important but comfort and functionality are the main matchwords.

Einige Entwürfe verstehen sich trotz ihrer Funktionalität auch als dekorative Elemente und bewegen sich auf dem schmalen Grat, welcher Skulpturen von Haushaltsgegenständen trennt. Obwohl die Ästhetik sehr wichtig ist, bleiben Gemütlichkeit und Funktionalität die elementaren Aspekte.

AVI table by Arne Verhoven. Magis Design

A simple, attractive table with its top in plywood and the frame in steel tube.

An inspirational workspace for inspiring endeavors. The Moontables' strong curves reveal a roomy inner storage space.

Certains designs, qui ont été conçus comme des éléments à la fois décoratifs et fonctionnels, se situent à la limite entre sculpture et object domestique.
Bien que l´esthetique de ces elements soit trés important, c´est avant tout la commodité et la fonctionnalité qui importe.

Algunos diseños se han concebido como elementos decorativos además de funcionales y se sitúan en la sutil línea que separa la escultura del objeto doméstico. La estética de estas piezas es muy importante, pero la comodidad es imprescindible.

Moontable Opaque
by Fuseproject

189

Air-tables *by Jasper Morrison. Magis Design*

Low air-tv table
by Jasper Morrison. Magis Design

Centomila tables *by James Irvine. Magis Design*

Magis Wagon
by Michel Young.
Magis Design

It's a coffee table, a container on four wheels in which to put magazines, books, candy. Other functions? Let's ask a child!

Kick Box by Robin Blatt & Cairn Young. Magis Design

A multi-use, stackable, easy-access container on wheels for home and office use.

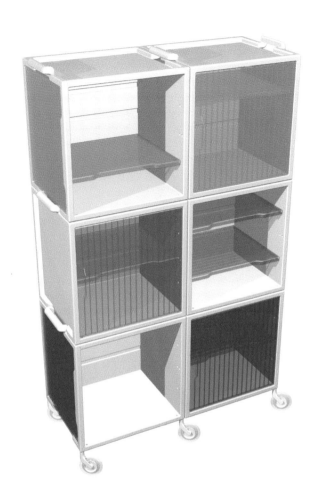

Kick Boxes
by Robin Blatt & Cairn Young.
Magis Design

A retractable wall table with a top in white werzalit.

Clino Once Again by Mario Mazzer. Magis Design

Desk Plus *by Stokke*

La Valise *by Ronan & Erwan Bouroullec. Magis Design*

195

Togo coat stand
by Enzo Mari. Magis Design

Slaphead children night light
by TKO Design

Superpatata
by Héctor Serrano

Pebbes by Marcel Wanders. Magis Design

Design by Lexon

Design by Jakob Wagner

Handtag
by Alex Lerm Design

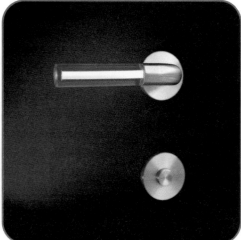

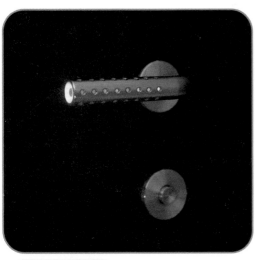

Handtag
by Alex Lerm Design

Handtag
by Alex Lerm Design

Home

Küche

Cocina

Kitchen

Cuisine

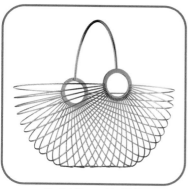

Basket Concept *by WMF*

Kitchen Utensils
These tools that we use everyday
require great precision in the design:
the materials have to guarantee
reliability, quality and durability, while
the form must be ergonomic.

Küchengeräte
Küchengeräte sind Werkzeuge des
täglichen Gebrauchs und müssen
sowohl bei der Wahl des Materials,
das Vertrauenswürdigkeit, Qualität
und Dauerhaftigkeit gewährleisten
muss, als auch in Bezug auf ihre
ergonomische Form im Design durch
ihre Präzision überzeugen.

Ustensiles de Cuisine
Ce sont nos outils quotidiens et,
comme tels, ils disposent d'un
design très précis tant pour la
recherche de matériaux assurant
leur fiabilité, qualité et longévité,
que pour leur forme ergonomique.

Utensilios de Cocina
Son nuestras herramientas de
uso diario y como tales poseen
un diseño muy preciso tanto en
la búsqueda de materiales que
aseguren su fiabilidad, calidad
y durabilidad, como en su forma
ergonómica.

Mill Chomargan *by WMF*

Perfect Plus *by WMF*

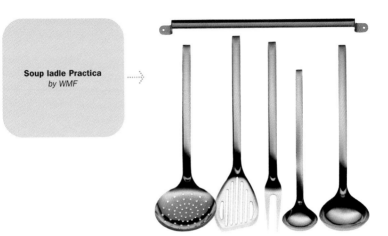

Soup ladle Practica *by WMF*

Creamer *by WMF*

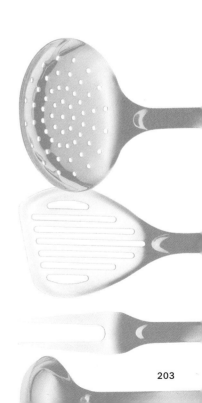

203

La Siesta by Alberto Martínez, Raky Martínez and Héctor Serrano

A fresh terracotta vessel for water, combining the look of a plastic bottle with the advantages of a traditional botijo. A different drinking experience.

Tic-tac
for Adhoc by Sol Design

Tic-tac measures the size of the egg and calculates the appropriate cooking time.

Ceramil Herb and Spice Mill by WMF

The spice mill appears in a design that is modern and cool. A grinding mechanism on the top stops crumbs from being scattered over the table when seasoning.

Don Pepino & Saltone by WMF

Poche à Eau water jug
by Radi Designers

This water jug is made of hand-blown glass. A simple pinch at the top of the jug and both, lip and handle, are formed.

Food Tool *for Zelco by Prime Studio Inc.*

A lightweight multipurpose culinary tool which combines full-size utensils in a compact and stylish package.

Float-tea lantern *by Forsythe + Mc Allen Design Associates*

Riva Cuttery *by Designit A/S*

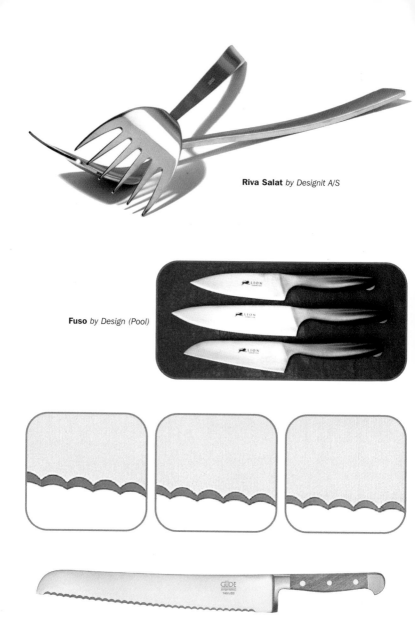

Riva Salat by Designit A/S

Fuso by Design (Pool)

Bread knife by Franz Güde GmbH

Pooto
by Laboratorio Pesaro S.r.l.

Ginger *by Paolo Pedrizzetti
& Associates*

Design by Büro für Form

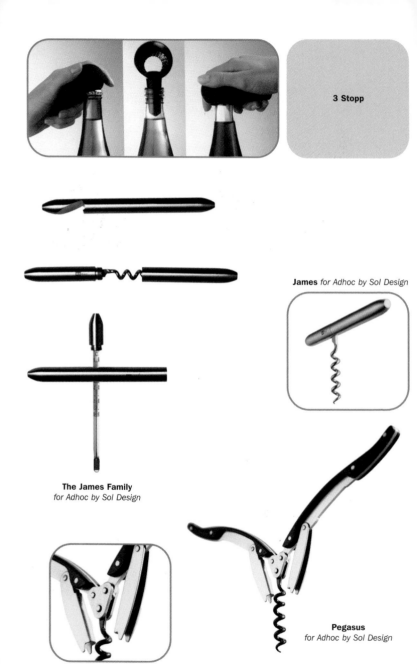

3 Stopp

James *for Adhoc by Sol Design*

The James Family
for Adhoc by Sol Design

Pegasus
for Adhoc by Sol Design

Wowo twist bowls by Smart Design

Wowo thermos by Smart Design

Wowo salad servers by Smart Design

Lush Lily trays
by Fuseproject

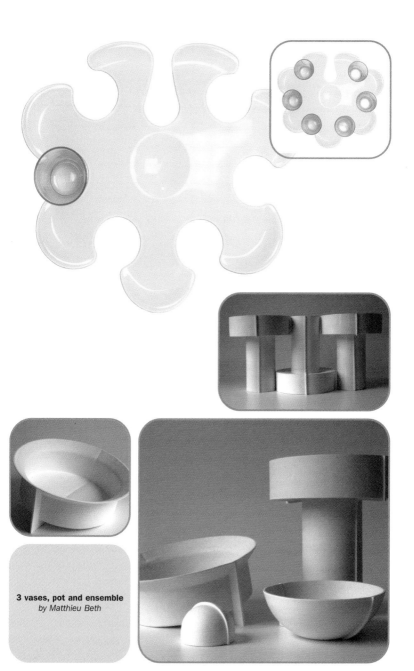

3 vases, pot and ensemble
by Matthieu Beth

213

Rat patrol kitchen timer
for Lec Inc. Japan by TKO Design

Kitchen timer
*for Lec Inc. Japan
by TKO Design*

Ahoi *for Koziol GmbH by Paolo Pedrizzetti & Associates*

Fruttifero
by Paolo Pedrizzetti & Associates

Fruit juicer

Kitchen

The kitchen has always been the heart of the home; nowadays it is also the room that most exploits the latest advances in technology, while design seeks to give them an attractive finish.

Küche

Die Küche war seit jeher das Zentrum jedes Haushalts und dort konzentrieren sich die meisten technologischen Neuerscheinungen. In diesem Zuge hat sich das Design auf die Verbesserung der dazugehörigen Gegenstände konzentriert.

Cuisine

Depuis toujours, la cuisine a été au cœur du foyer et a concentré le plus grand nombre d'avancées technologiques. Suivant celles-ci, le design s'est porté sur l'amélioration des objets les accompagnant.

Cocina

La cocina es el corazón del hogar y donde se han concentrado el mayor número de avances tecnológicos. A remolque de éstos, el diseño se ha concentrado en mejorar los objetos que los acompañan.

Microwave
for Samsung by Ziba

Uplift teakettle *by Oxo*

Café Elite *by Princess*

Gribus *by Klaus Hackl. Magis Design*

Angled measuring cup
by Oxo

Rubbermaid waste cans
by Ziba

Storage and serving system
by WMF

Tupperware-Vitalic cookware range *by Enthoven Associates*

Kitchen Accessories
Many of today's accessories complement utility with humor and playfulness.

Küchenzubehör
Das meiste Zubehör verbindet Nützlichkeit mit Humor und Spielerei.

Accessoires de Cuisine
La majeure partie des accessoires conjuguent utilité avec humour et jeu.

Accesorios de Cocina
La mayoría de los de los accesorios combinan utilidad, con humor y juego.

Vallauris Robot Family
by Florence Doleac.
Radi Designers

Peppermill and saltshaker
by Oxo

Design by Oxo

221

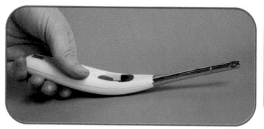

Flash gas highler
by Andries Van Onck & C.s.n.c.

Mauscan opener
by Animal Design Factory

Spinner
by Animal Design Factory

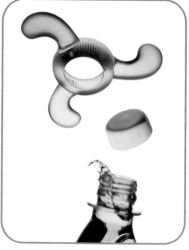

Spaghetti utensils
for Authentics by Laurence + Constantin Boym

Wet dry bath caddy
for Authentics by Laurence + Constantin Boym

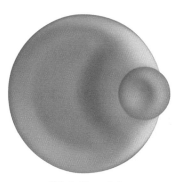

Chip 'n Dip party tray
for Authentics by Boym Design Studio

Salvation
by Laurence + Constantin Boym

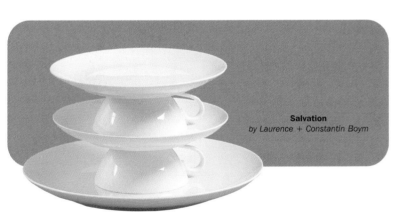

Salvation
by Laurence + Constantin Boym

Salvation
by Laurence + Constantin Boym

Step 90 *by Magis Design*

Mago *by Magis Design*

Paper table
for DMD by Constantin Boym

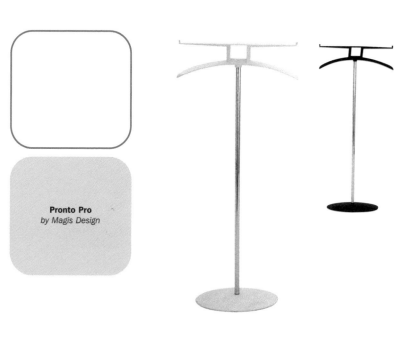

Pronto Pro
by Magis Design

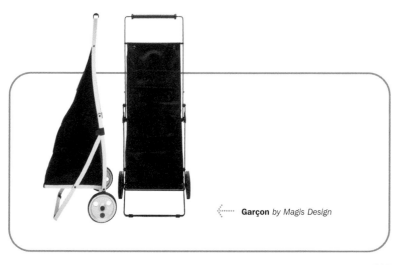

Garçon *by Magis Design*

Stepladder Kartell *by Andries Van Onck & C.s.n.c.*

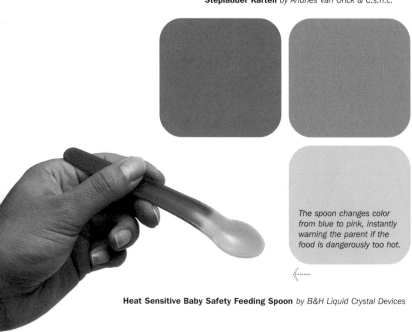

The spoon changes color from blue to pink, instantly warning the parent if the food is dangerously too hot.

Heat Sensitive Baby Safety Feeding Spoon *by B&H Liquid Crystal Devices*

Pipe Dreams *by Magis Design*

Dish Doctor *by Marc Newson. Magis Design*

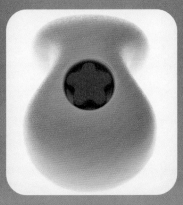

Rock *by Magis Design*

Lilliput Salt and Pepper
*for Alessi
by Stefano Giovannoni*

Bruce table highler
for Alessi by Stefano Giovannoni

Pino funnel
*for Alessi by
Stefano Giovannoni
and Miriam Mirri*

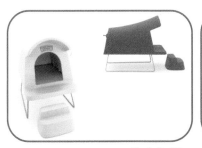 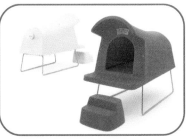

Magis Dog House by Magis Design

Pedal Bowl by Pedal Bowl Products Ltd.

Tapaspinner *by Jakob Wagner*

Crockery
The growing interest in traditional materials such as pottery, porcelain and glass has resulted in highly sinuous objects with smooth forms.

Geschirr
Die zunehmende Beschäftigung mit traditionellen Materialien wie Keramik, Porzellan oder Glas ließ Objekte mit sanfteren und geschwungeneren Linien entstehen.

Vaisselle
Les recherches croissantes dans les matériaux traditionnels, comme la céramique, la porcelaine ou le verre, ont engendré des objets plus sinueux et aux formes plus douces.

Vajillas
La creciente investigación en materiales tradicionales como la cerámica, la porcelana o el vidrio han dado de sí objetos mucho más sinuosos y de formas más suaves.

Table Dishes *by Jakob Wagner*

Table Dishes *by Jakob Wagner*

Table Dishes *by Jakob Wagner*

Foilcutter *by Jakob Wagner*

Korkscrewer *by Jakob Wagner*

Vacuumer by Jakob Wagner

235

Home

Badezimmer

Salle de Bain

Baño

Bathroom

Wash-basin
for Villeroy & Boch by Frog Design

Dreamscape wash-basin ensemble
by Michael Graves & Associates

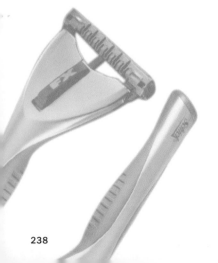

Bathrooms
The bathroom has forsaken the coldness and sterility of the past to become a space where design and technological innovation combine to provide comfort and convenience.

Badezimmer
Das kalte und sterile Badezimmer wurde ersetzt durch Räume, in denen Design und technologische Neuerungen alle Funktionen auf leichte und komfortable Art möglich machen.

Tangible *for Niethammer by Sol Design*

Bathroom
for Villeroy & Boch by Frog Design

Organizer *by Paolo Pedrizzetti & Associates*

Salle de Bain
La salle de bain a cessé d'être une pièce froide et aseptisée pour se transformer en un des lieux ou le design et les innovations technologiques permettent de réaliser toutes les fonctions facilement et confortablement.

Baño
El baño ha dejado de ser la estancia fría y aséptica para convertirse en habitaciones donde el diseño y las innovaciones tecnológicas permiten realizar todas las funciones de una manera fácil y confortable.

Bable *by Paolo Pedrizzetti & Associates*

Showers and Sinks
These are obviously still the key elements of a bathroom but they have acquired a greater formal complexity. Technological innovations have introduced a new concept in showers.

Sanitäre Anlagen
Sie spielen, allerdings in neuer formaler Vielgestaltigkeit, immer noch die Hauptrolle im Badezimmer. Der Schwerpunkt technologischer Entwicklungen liegt dabei auf neuen Konzepten für die Dusche.

Bable 2 *by Paolo Pedrizzetti & Associates*

Girad *by Sol Design*

Nautilus shower platform
*by Paolo Pedrizzetti &
Associates*

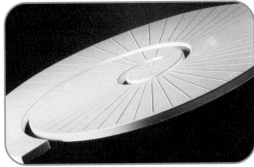

Sanitaires
Ils demeurent un des éléments clés de la salle de bain mais ont gagné en
complexité formelle. Les innovations technologiques portent essentiellement sur
un nouveau concept de douche.

Sanitarios
Continúan siendo las piezas claves del baño pero han ganado en complejidad
formal. Las innovaciones tecnológicas se centran en un nuevo concepto de
ducha.

241

Toilet brush & plunger
by Oxo

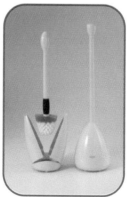

Toilet brush & plunger
by Oxo

243

244

American safety razor
by Henry Dreyfuss Associates

Schick FX *for Wilkinson*

245

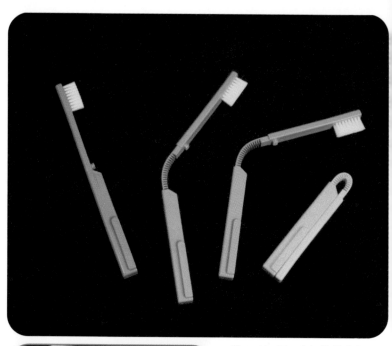

Manual toothbrush by Emilio Ambasz +
Associates Inc

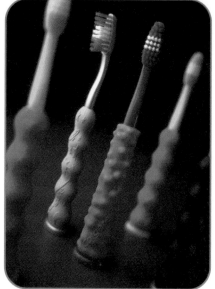

**Colgate Grip'ems
toothbrush**
by Ecco

Gripper toothbrush
for Oral B by Ideo Europe

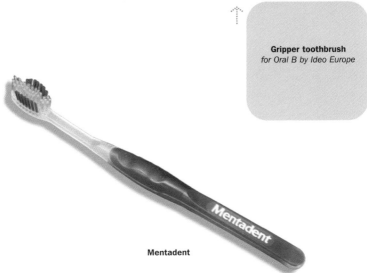

Mentadent

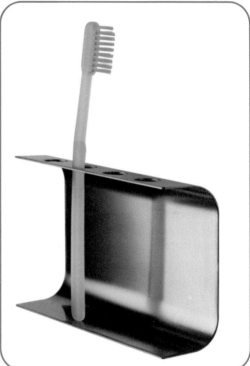

Toothbrush rack
by Almén:gest Design

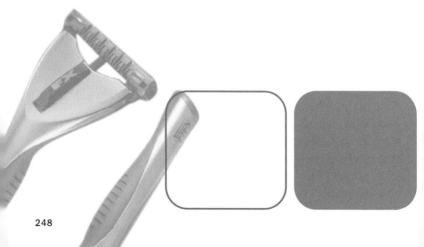

248

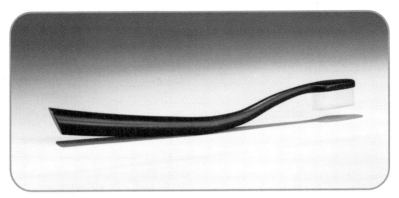

Design by Emilio Ambasz + Associates Inc.

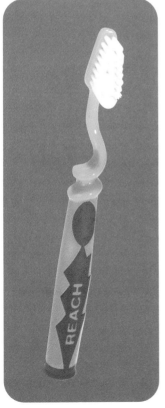

Design by Emilio Ambasz + Associates Inc.

J & J Wondergrip toothbrush *by Smart Design*

Oxo Suction Cup Soap Dishes by *Smart Design*

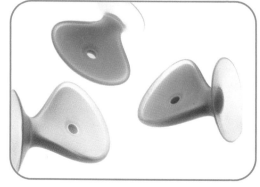

Oxo Suction Cup Mirror
by Smart Design

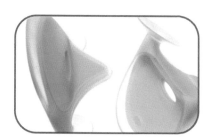

Shower Radio *by Design (Pool)*

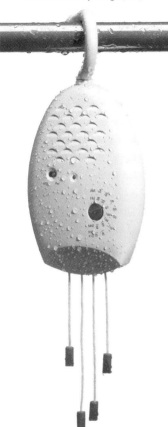

Bathroom Accessories
Surprise and elegance are two prominent features of today's designs for bathroom accessories, helping to give the space a personal touch.

Zubehör für das Badezimmer
Bei der persönlichen Gestaltung des Badezimmers geht die Tendenz des Designs ganz klar zu Eleganz und Überraschungseffekten beim Zubehör.

Accessoires de Bain
Afin de personnaliser la pièce, l'élégance et la surprise sont les deux tendances les plus claires du design des accessoires de bain.

Accesorios de Baño
Para personificar la habitación, la elegancia y la sorpresa son las dos tendencias más claras en el diseño de accesorios de baño.

Climber
by Paolo Pedrizzetti & Associates

251

Design by Lexon

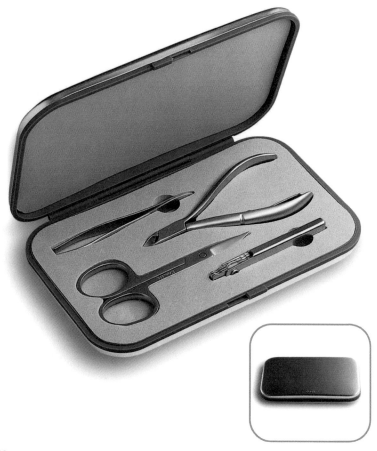

Soap dispender
for Dial by Ziba

Meeting the needs of a wide variety of restroom users, this dispender is activated simply by placing one hand under the dispender and depressing the oversized lever.

253

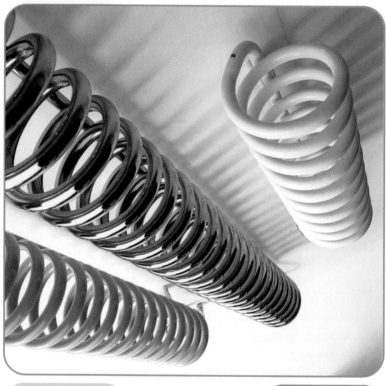

Hot Spring Radiator
by Priestman Goode

Squeegee
for Cleret by Ziba

The design continues to set the standard for showering: squeegees!

Pipinette *by Pipinette AB*

Merdorino toilet brush
for Alessi
by Stefano Giovannoni

Regatta *by Hoesch Design*

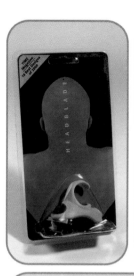

A lot of people shave their head ...

It's time someone designed a razor for them.

Head Blade
*by The Head Blade
Company, LLC*

259

Bathroom
for Hoesch
by Massimo Iosa Ghini

Bathroom
for Duravit
by Massimo Iosa Ghini

Bathroom
for Duravit
by Massimo Iosa Ghini

Toilets
The major companies are still finding new ways in which to gradually improve our personal hygiene.

Toiletten
Die großen Firmen übertreffen sich mit ihren Angeboten an neuen Modellen zur Verbesserung unserer hygienischen Gepflogenheiten.

Toilettes
Les grandes entreprises proposent continûment de nouvelles innovations qui améliorent peu à peu nos habitudes d'hygiène.

Aseo
Las grandes empresas continúan ofreciendo nuevas innovaciones que van mejorando poco a poco nuestros hábitos de limpieza.

Corner bath with apron
for Hoesch by Massimo Iosa Ghini

Built-in whirlpool
for Hoesch by Massimo Iosa Ghini

Booth with apron
for Hoesch by Massimo Iosa Ghini

Shower tray with Cleves shower partition
for Hoesch by Massimo Iosa Ghini

**Furniture washbasin with vanity unit
mirrorcabinet** *for Duravit
by Massimo Iosa Ghini*

Self Sceen
for Duravit by Massimo Iosa Ghini

Floor-standing WC and bidet
for Duravit by Massimo Iosa Ghini

Faucets
The latest designs are not only concerned with formal aspects but also aim to be ecological and ergonomic.

Armaturen
Hier werden nicht nur die formalen Kriterien berücksichtigt, sondern auch die Lösung ökologischer und ergonomischer Probleme.

Robinetterie
Des pièces sont développées non seulement en gardant les aspects formels à l'esprit, mais également afin de résoudre des questions écologiques et ergonomiques.

Soap dispender
for Duravit/Dornbach
by Massimo Iosa Ghini

Towel ring
for Duravit/Dornbach
by Massimo Iosa Ghini

Soap dish
for Duravit/Dornbach
by Massimo Iosa Ghini

**Single lever washbasin
mixer** *for Dornbach*
by Massimo Iosa Ghini

Hole washbasin tap fitting
for Dornbach
by Massimo Iosa Ghini

Grifería
Se están desarrollando
piezas no tan solo
preocupadas en
aspectos formales sino
también en solucionar
problemas ecológicos y
ergonómicos.

Hole bath fitting *for Dornbach by Massimo Iosa Ghini*

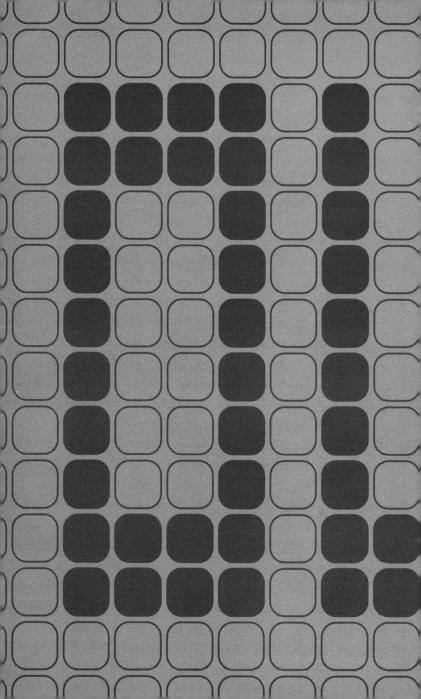

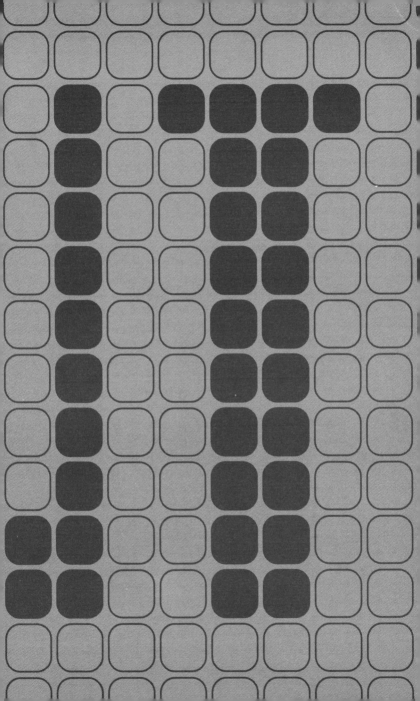

Complements

Accessoires

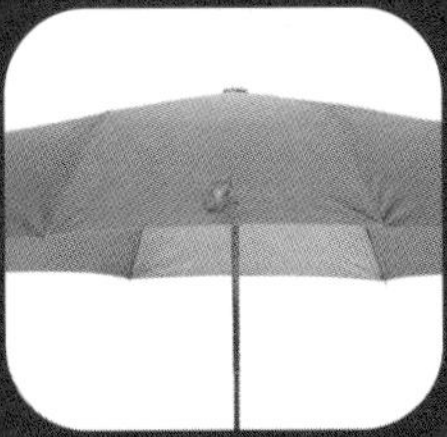

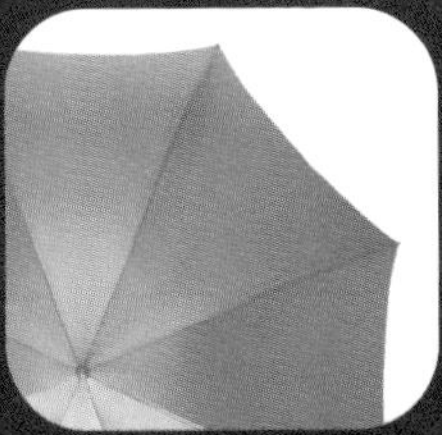

Zubehör

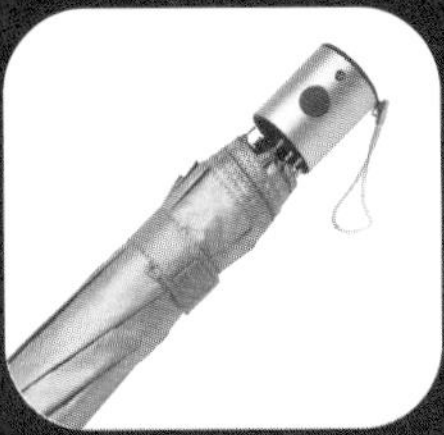

Complementos

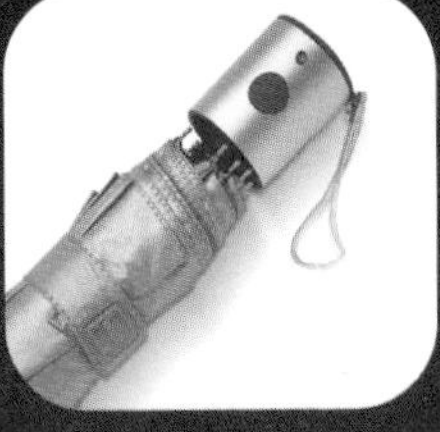

Monkey Alpha Cubic
by Design (Pool)

Revo sunglasses
by Ecco

Ecco identifies unique design criteria from comfort, slip, adjustability to peripheral vision to lens characteristic.

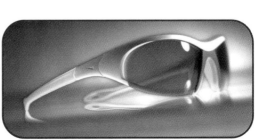

Ruiz sunglasses
for Nike by Ideo Europe

269

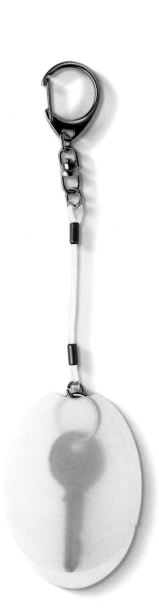

Design by Lexon

Complements

The advance of technology leads to greater speed and power and more performance qualities. Objects are also getting smaller–this is the so-called miniaturization process, which is bringing progress into our everyday lives.

Design by Lexon

Zubehör
Parallel zu den Neuerungen der Technologie, die größere Geschwindigkeiten, bessere Leistungen und mehr Möglichkeiten mit sich bringen, reduziert sich die Größe der Objekte zunehmend. Dieser Prozess der Miniaturisierung erlaubt es, den Fortschritt immer stärker in unser tägliches Leben zu integrieren.

Compléments
Parallèlement aux avancées technologiques – plus de vitesse, de puissance et des prestations plus nombreuses – le volume des objets se réduit. C'est le processus maintenant reconnu de miniaturisation qui permet d'incorporer des avancées chaque jour plus proches de nous.

Complementos
Paralelamente al avance de la tecnología –mayor rapidez, más potencia y mayores prestaciones– se está reduciendo el volumen de los objetos. Es el llamado proceso de miniaturización, que está permitiendo incorporar los avances cada vez más cerca de nosotros.

Design by Lexon

Habillement

Fonctionnalité, beauté, précision, utilité, commodité et facteur émotionnel du produit sont des caractéristiques définissant un bon design dans ce domaine. Il convient également de mettre en avant le design sportif et sa recherche sur les matériaux qui ont eu une influence profonde dans le secteur de la mode.

Vestir

Funcionalidad, belleza, precisión, utilidad, comodidad y el factor emocional del producto son características que definen un buen diseño en este campo. También apuntar que el diseño deportivo y su investigación de materiales han tenido una gran influencia en el sector de la moda.

Peau d'housse by Florence Doléac & Gaspart Yurkivich. Radi Designers

Clothes

Functionality, beauty, precision, utility, comfort and an emotional element are the characteristics that define a good design in this field. It must be added that sports design, with its research in new materials, has had a great influence on the world of fashion.

Kleidung

Ein gutes Design auf diesem Gebiet zeichnet sich durch Funktionalität, Schönheit, Präzision, Nützlichkeit, Bequemlichkeit und den emotionalen Faktor des Produktes aus. Dabei muss betont werden, dass das Sportmoden-Design und die damit verbundenen Materialentwichlungen den Modesektor stark beeinflusst haben.

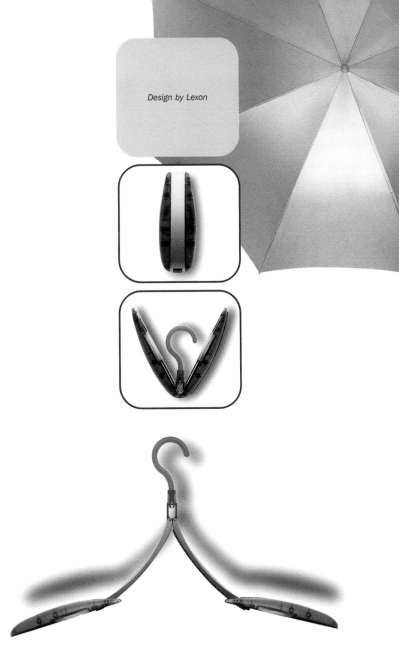

Design by Lexon

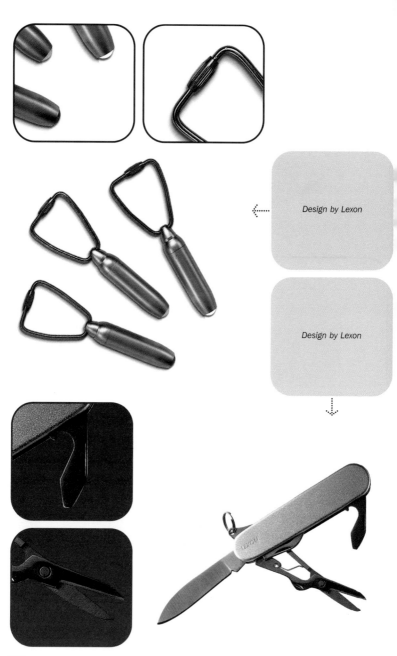

Design by Lexon

Design by Lexon

Design by Lexon

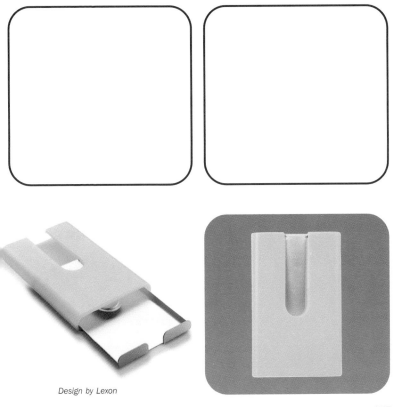

Design by Lexon

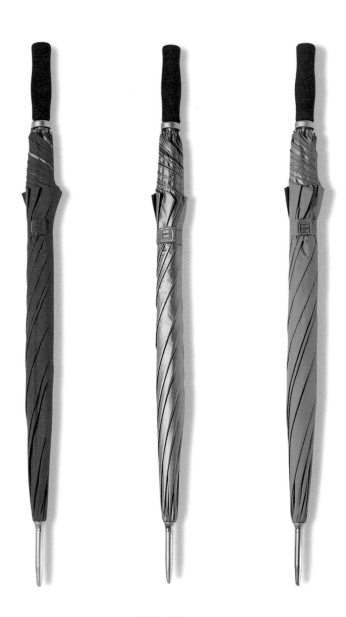

Design by Lexon

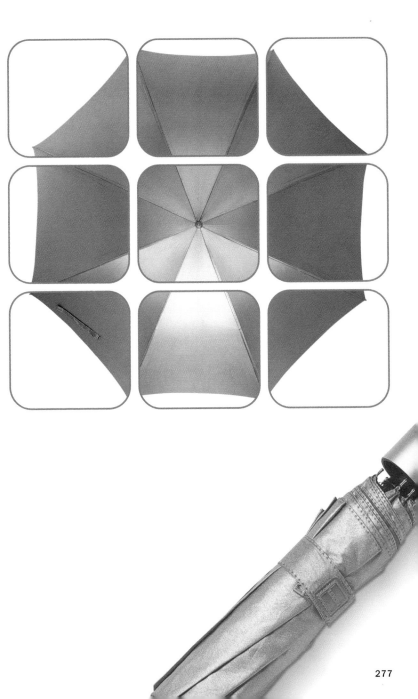

Design by Lexon

Design by Lexon

Slam by Büro für Form

Electronics
Most experimental design or design utopias are intent on studying the possibilities offered by miniaturization.

Elektronische Geräte
Ein Großteil des experimentellen oder utopischen Designs richtet sich nach den Möglichkeiten, welche die Miniaturisierung dabei zulässt.

Électronique
La plupart du design expérimental, ou utopies de design, prend la voie de l'étude des possibilités offertes par la miniaturisation.

Electrónicos
La mayoría del diseño experimental o utopías del diseño se encaminan al estudio de las posibilidades que ofrece la miniaturización.

Soft Telephone by Emilio Ambagz + Associates Inc.

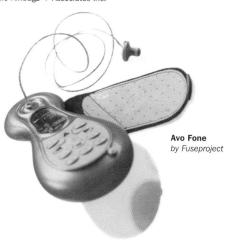

Avo Fone
by Fuseproject

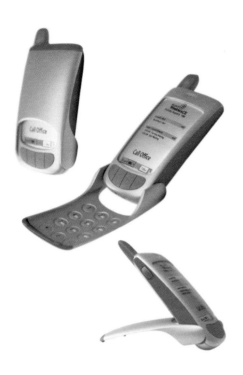

CE PDA Phones
for Microsoft by Fuseproject

CE PDA Phones
for Microsoft
by Fuseproject

Three phones for Microsoft, with designs fusing cell phones and personal digital assistance.

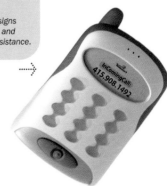

The outercover slides, flips and rotates, revealing larger screens for e-mail and web browsing, sliding back to protect the screen while in standard phone function.

CE PDA Phones
for Microsoft by Fuseproject   ......>

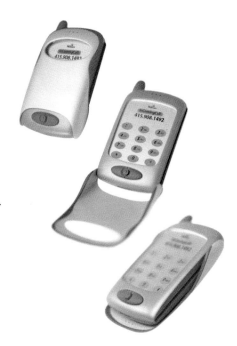

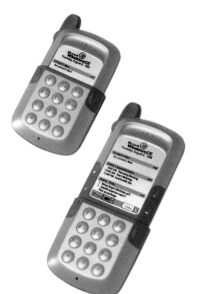

<....... **CE PDA Phones**
for Microsoft by Fuseproject

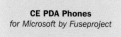

Design by Lexon

Design by Lexon

Design by Lexon

Design by Lexon

284

Motorola Timex Beepwear
by Ecco

The merging of two extraordinary brands, Motorola and Timex, has brought out Beepwear, the union of pager technology in a watch.

Design by Lexon

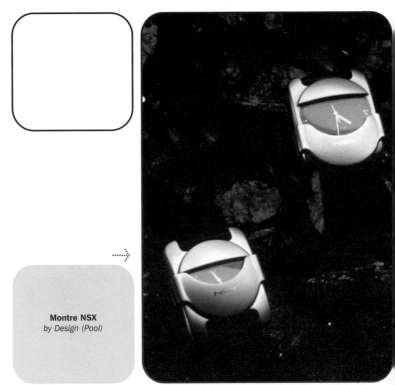

Montre NSX
by Design (Pool)

Watches *for Timberland by Smart Design*

Design by Lexon

Design by Lexon

288

Design by Lexon

Design by Lexon

Design by Lexon

289

Business Class Briefcase
by Radi Designers

Accessories
The large-scale application of state-of-the-art materials makes it possible to redesign old products by seeking a new harmony between their form and their use.

Zubehör
Die umfangreiche Verwendung von Materialien der letzten Generation ermöglicht es, alte Produkte auf der Suche nach einer neuen Harmonie zwischen Form und Gebrauch neu zu gestalten.

Accessoires
L'application massive des matériaux de dernière génération permet de concevoir à nouveau d'anciens produits en quête d'une nouvelle harmonie entre la forme et l'usage.

Accesorios
La masiva aplicación de materiales de última generación permite rediseñar antiguos productos en busca de una nueva armonía entre su forma y uso.

Design by Lexon

GripIt Shopping Bag Handle
for Yagi by Ziba

*An innovative and ergonomic
solution for a comfortable
shopping tour.*

Design by Lexon

Design by Lexon

Design by Lexon

Icon Suitcase *by Opius Limited*

Concept de Superior
by Design (Pool)

XForm *by Design (Pool)*

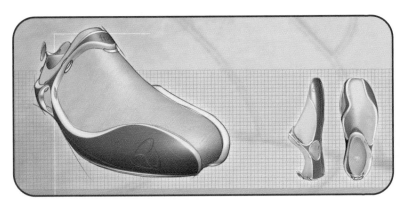

Prana future footware concept *by Frog Design*

XForm *by Design (Pool)*

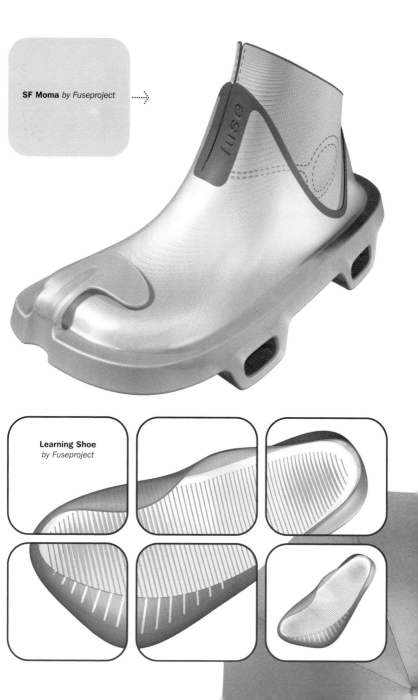

SF Moma *by Fuseproject*

Learning Shoe
by Fuseproject

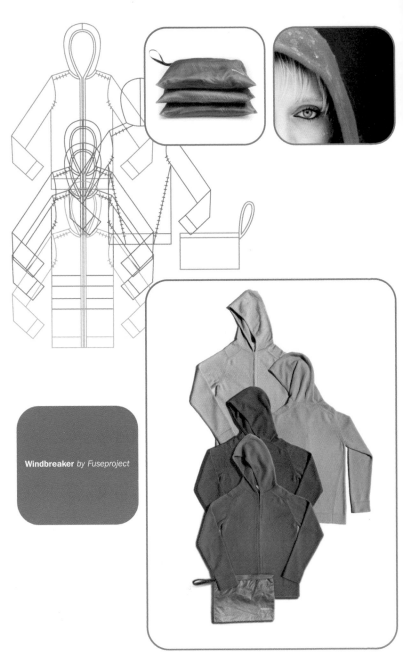

Windbreaker *by Fuseproject*

Fusekin *by Fuseproject*

Sport
by Bézenville
Paris Urbain

Design by Lexon

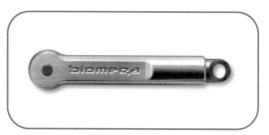

Biomega *by Jakob Wagner*

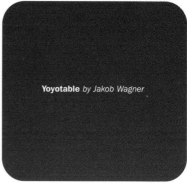

Yoyotable *by Jakob Wagner*

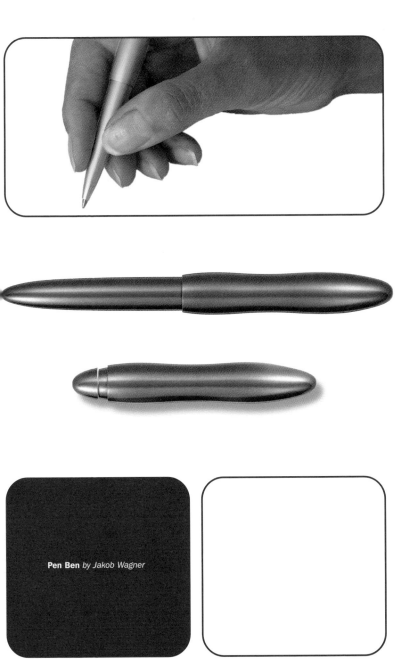

Pen Ben *by Jakob Wagner*

Urban
Furniture

Mobiliario Urbano

Urbanes Mobiliar

Mobilier Urbain

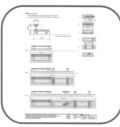

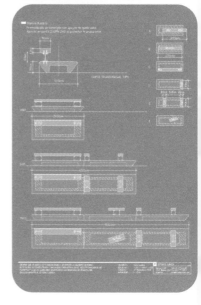

Banco Rambla
by Diana Cabeza

**Hand sketches for
Banco Rambla**
by Diana Cabeza

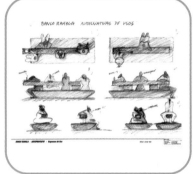

Lungo Mare
by Escofet

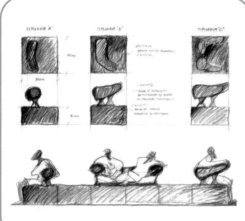

Picapiedras
by Diana Cabeza

Whippet Bench *by Radi Designers*

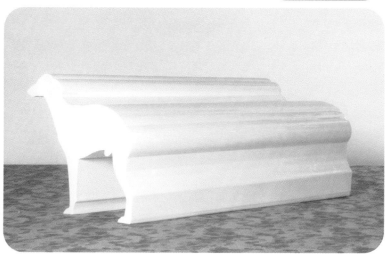

Europa bench *for Vesco by Designit A/S*

Urban Furniture
The public spaces around us are packed with objects that provide us with support, assistance or information. They form part of the urban landscape and we often only notice them when we arrive in a new city, as a good urban design must adapt to cultural differences.

Urbanes Mobiliar
Der öffentliche Raum, in dem wir uns bewegen, ist voll von Einrichtungen, die uns beinflussen, uns helfen oder informieren. Sie gehören zum Stadtbild und oft bemerken wir sie erst, wenn wir in eine andere Stadt ziehen. Ein gutes Stadtdesign zeichnet sich dadurch aus, dass es den kulturellen Unterschieden gerecht wird.

Mobilier Urbain
L'espace public qui nous entoure est rempli d'objets qui nous assistent, nous aident et nous informent. Ils font partie du paysage urbain et, souvent, nous n'en prenons conscience que lorsque nous changeons de ville, un bon design urbain devant s'adapter aux spécificités culturelles.

Mobiliario Urbano
El espacio público que nos rodea está lleno de objetos que nos asisten, nos ayudan o nos informan. Forman parte del paisaje urbano y muchas veces solamente nos fijamos en ellos cuando cambiamos de ciudad, pues un buen diseño urbano ha de adecuarse a las diferencias culturales.

Petra (Banco Pausa)
by Manufacturas Mago SL

Enthoven line corner roof – drain *by Enthoven Associates*

Aseda

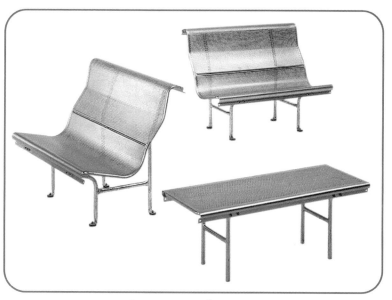

Banco Catalano *by Óscar Tusquets*

Banco Catalano
by Óscar Tusquets

Elehna *by Nusser*

RPI Wave Bench
by Ecco

Limbank

Reposa

Torento

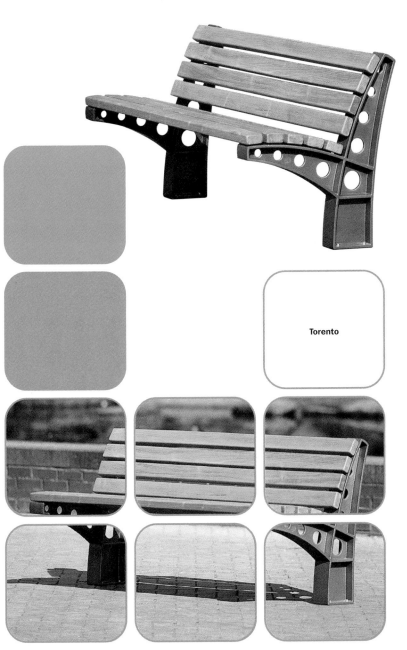

Torento

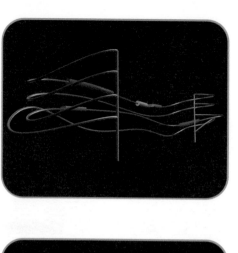

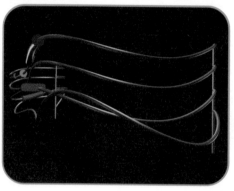

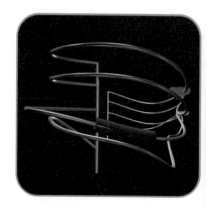

Brainstorm by Terminal NYC Brainstorm

**Clear Channel Adshel –
Enthoven line trainshelter
with public phone**
by Enthoven Associates

**Clear Channel Adshel –
Enthoven line trainshelter**
by Enthoven Associates

Furnishings

Two points must be born in mind in the design of urban furnishings. The aesthetic impact sought and the search for materials that can endure all types of weather.

Mobiliar

Zwei Aspekte müssen beim Design des städtischen Mobiliars berücksichtigt werden: die beabsichtigte ästhetische Wirkung und das zu verwendende Material, das allen Arten von Witterungs-einflüssen Stand halten muss.

Mobilier

Deux points doivent être pris en compte pour le design du mobilier urbain. L'impact esthétique désiré et l'étude du matériau à utiliser afin de faire face à tout type d'intempérie.

Mobiliario

Hay dos puntos a tener en cuenta en el diseño de mobiliario urbano: el impacto estético deseado y el estudio del material a utilizar para soportar todo tipo de inclemencias.

Parada by King & Miranda Design

Cemusa by King & Miranda Design

319

Mairie Mobile
by J.C. Decaux. Design (Pool)

Green Light
by Harrit & Sørensen a/s

321

Aspapel by Mormedi

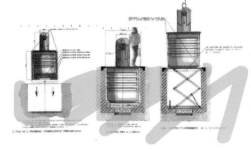

Allibert by Design (Pool)

Allibert Development Urbain-France is a container project for collecting and selecting urban garbage.

Power Pylon *by Bystrup*

Rama *by Santa & Cole*

Via Farola *by Santa & Cole*

Coll

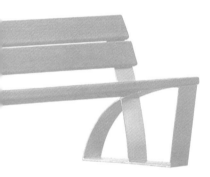

Coll

Joriano *by Andries Van Onck & C.s.n.c.*

Viacom *for Giraudy by Design (Pool)*

Arcomat by Design (Pool)

Flash Vote *by Design (Pool)*

eCast Internet Jukebox Gaming Device
by Frog Design

Information Points
This field of industrial design must strive for accessibility, ease in handling, clarity and robustness.

Informationsstationen
Auf diesem Gebiet muss das industrielle Design Zugänglichkeit, leichte Handhabung, gute Bildqualität und Widerstandsfähigkeit bieten können.

Points d'Information
Accessibilité, usage facile, clarté, robustesse sont des caractéristiques indispensables dans ce domaine du design industriel.

Puntos de Información
Accesibilidad, fácil manejo, claridad y robustez son características que ha poseer este campo del diseño industrial.

Web Station *for Street Space by Ideo Europe*

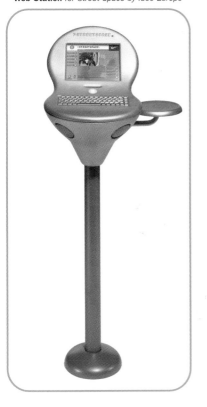

**Fountaine de l'An 2000
drinking fountain for Paris**
by Radi Designers

*The fountain is made from
cast steel. It's generated
by a gyratory extrusion of
two female profiles.
Frozen in this movement,
the arm and hand are
outstretched to offer water
to the passer-by.*

Fountaine de l'An 2000
drinking fountain for Paris
by Radi Designers

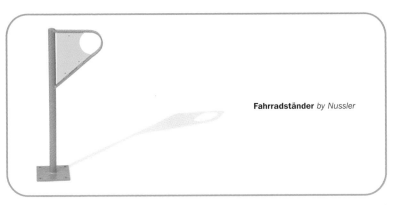

Fahrradständer *by Nussler*

Signum

Indico

Indico

Signum

333

Clasic

Limes

Metro

Metropol

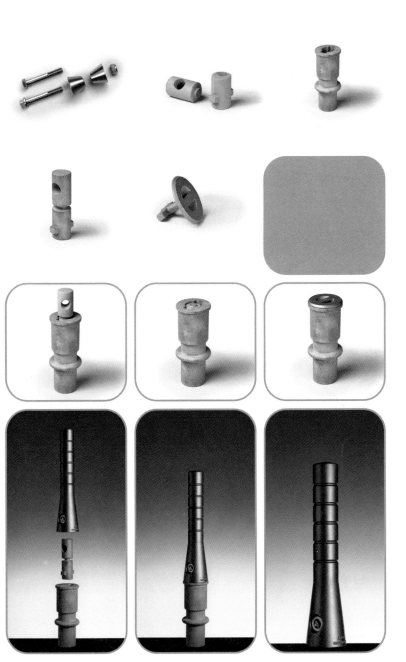

Leisure

Ocio

Freizeit

Loisirs

Scoot's hydrogen fuel motor propels the user along without harming the environment. Fold down in seconds for easy carrying and placing.

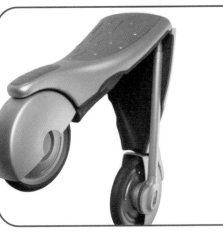

Leisure

Industrial design must take into account various factors (economic, functional, technological, social, etc.) but it must not overlook its symbolic function. The interaction between an object and a spectator is just as important as any other, and this is perhaps most noticeable in the field of leisure.

Freizeit

Obwohl das industrielle Design verschiedene Faktoren zu berücksichtigen hat (wirtschaftliche, funktionelle, technologische, gesellschaftliche...), darf es jedoch die symbolische Funktion nicht außer Acht lassen. Die notwendige und sehr wichtige Kommunikation zwischen Objekt und Individuum kann vielleicht auf dem Gebiet der Freizeit am besten gewürdigt werden.

Loisirs

Le design industriel doit tenir compte de divers facteurs (économiques, fonctionnels, technologiques, sociaux…) mais sans oublier la fonction symbolique. La communication qui doit exister entre l'objet et l'individu est aussi importante, et c'est peut-être dans le domaine des loisirs qu'elle s'apprécie le mieux.

Ocio

El diseño industrial debe tener en cuenta diversos factores (económicos, funcionales, tecnológicos, sociales…), pero no debe olvidar la función simbólica. La comunicación que ha de haber entre el objeto y el individuo es igual de importante, y es quizás en el campo del ocio donde esto se puede apreciar más.

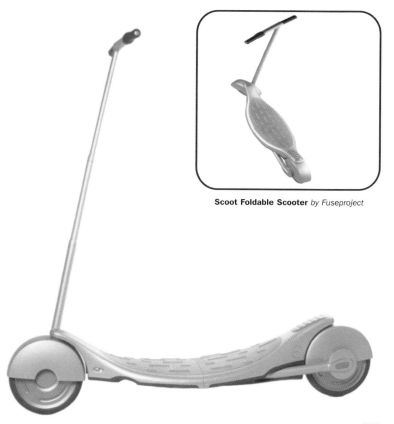

Scoot Foldable Scooter *by Fuseproject*

Come Bike by Design (Pool)

Citybike *by Lexon*

Design by Lexon

Design by Lexon

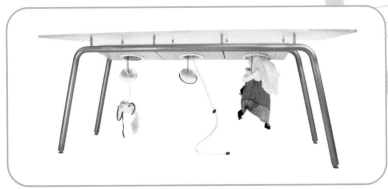

Undercover Table *by Thom Faulders. Beige Architecture Design*

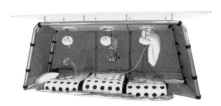

Dali Evidence 470 Top *by Designit A/S*

Dual Stereo System *by Frog Design*

Stick Radio

Creation of a companys' brand identity through creative convergence: two interlooking 'J's that form a globe.

Javad Positioning System *by Frog Design*

Pre'view Camera
for Sanrus
by Prime Studio Inc.

Olympus ·····>

PopShot *for Polaroid by Ideo Europe*

Polaroid Vision
by Henry Dreyfuss Associates

APS Cameras *for Fuji by Ziba*

Olympus Camera
by Frog Design

Polaroid Spectra
by Henry Dreyfuss Associates

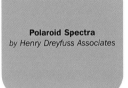

Polaroid

347

Tennis Racket for Head *by Frog Design*

Dikar *by Tres D*

Sport

In products aimed at the general public, the first application of technical innovations and research into the latest materials is often seen in sports design.

Sport

Technischen Neuerungen und neuste Materialentwicklungen finden ihre erste Anwendung häufig bei Produkten aus dem Bereich des Sportdesigns, die für ein großes Publikum bestimmt sind.

Sports

Les innovations techniques et l'étude des matériaux les plus avancés trouvent souvent leurs premières applications dans des produits destinés au grand public, dans le domaine du design sportif.

Deportes

Las innovaciones técnicas y el estudio de los materiales más punteros suelen encontrar sus primeras aplicaciones en productos destinados al gran público en el campo del diseño deportivo.

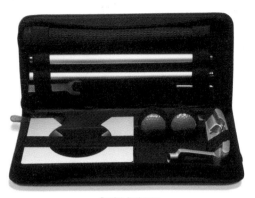

Design by Lexon

Tkoti Urban Hybrid Bicycle
by TKO Design

Unique design where material, processes and dynamic meet to produce an ultra-light beautiful simple urban bike.

Mountain bike *by Lexon*

<------- **MooBoo All Road Bike**
for Kildemoes by Designit A/S

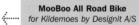

351

Design by Lexon

Impacy

Design by Lexon

Design by Lexon

Design by Lexon

353

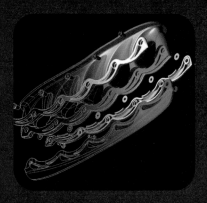

Würtner t'blade *by Frog Design*

Snowboard Binding Z1 *for Morrow by Ziba*

Fenwick Spinning Fishing Real *by Ziba*

Berkley *by Ideo Europe*

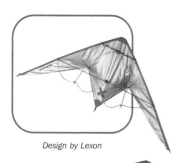

Design by Lexon

Hobbies
These are not just objects designed to fulfill a specific function, they also connect us to pastimes we enjoy.

Hobbys
Die gestalteten Objekte müssen nicht nur eine spezifische Funktion erfüllen, sondern auch eine Verbindung mit der Welt herstellen, an der wir Gefallen haben.

Hobbies
Ce ne sont pas uniquement des objets conçus pour remplir une fonction spécifique, mais aussi pour établir une relation avec un monde qui nous attire.

Hobbies
No son sólo objetos diseñados para cumplir una función específica; son objetos que nos relacionan con el mundo que nos gusta.

Design by Lexon

357

Design by Lexon

Design by Lexon

Design by Lexon

Design by Lexon

359

Terracottem by Pilipili Product Design

Savic Seedy by Pilipili Product Design

Powerflower by Designpartners

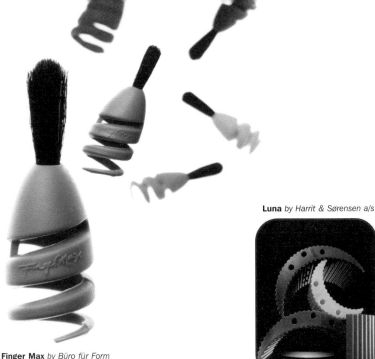

Luna by Harrit & Sørensen a/s

Finger Max by Büro für Form

Children
The parameters of a good design for children are slightly different from those of designs for adults. The colors have to be bright to attract attention, and the surfaces and the materials used must be totally harmless.

Kinder
Die Parameter eines guten Designs für Kinder unterscheiden sich etwas von denjenigen eines Designs für Erwachsene. Die Farben müssen lebhaft sein, um Aufmerksamkeit zu erregen, und die verwendeten Materialien vollkommen unschädlich.

Enfants
Les paramètres d'un bon design pour les enfants sont légèrement différents de ceux pour les adultes. Les couleurs doivent être vives afin d'attirer l'attention et les surfaces et matériaux employés totalement inoffensifs.

Infantil
Los parámetros de un buen diseño para niños son ligeramente diferentes que los diseños para adultos. Los colores han de ser vivos para llamar la atención y las superficies y materiales empleados han de ser totalmente inocuos.

Diam Bike
by Design (Pool)

Conception of a bike based on plastic-recycling technology.

"Gogo" Creative Learning
by Harrit & Sørensen a/s

Thrustmaster Gaming Device Frog Master
by Ziba

Gravis Exterminator Joystick
by Ecco

Wingman Formula Force
for Logitech by Ideo Europe

Design by Jakob Wagner

Design by Jakob Wagner

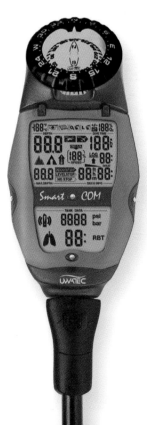

Smartcom Console *by Jakob Wagner*

PSS 500 *by Jakob Wagner*

Troikako by Jakob Wagner

Neverlos by Jakob Wagner

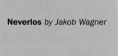

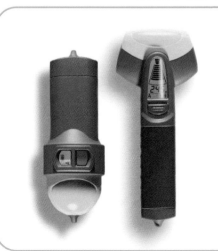

Skubabro
by Jakob Wagner

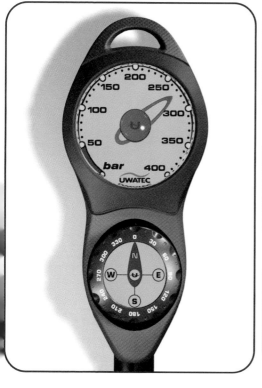

Lineto *by Jakob Wagner*

367

Vehicles

Véhicules

Vehículos

Fahrzeuge

Galaxy exterior *by Henry Dreyfuss Associates*

Galaxy interior *by Henry Dreyfuss Associates*

Vehicles

The most popular field in industrial design is undoubtedly the design of vehicles. It is in this type of product that the three basic functions of design are most clearly visible: the functional, the indicative and the symbolic.

Fahrzeuge

Der beliebteste Sektor des industriellen Designs ist sicherlich das Design von Fahrzeugen. Bei Fahrzeugen sind die drei Grundfunktionen des Designs am deutlichsten zu erkennen: Funktionalität, Bedeutung und Symbolik.

Véhicules

Sans conteste, la création automobile constitue le champ du design industriel le plus prisé. Ce sont ces types de produits qui reflètent le plus clairement les trois fonctions de base du design : fonctionnelle, indicative et symbolique.

Vehículos

Seguramente el campo más popular del diseño industrial es el diseño de vehículos. Es en este tipo de productos donde las tres funciones básicas del diseño se ven más claramente: la funcional, la indicativa y la simbólica.

Falcon Jet *by Henry Dreyfuss Associates*

Concept Car Zoom
by Renault

Concept Car Ludo
by Renault

Asymetric doors offer easy access, while the aluminum structure is displayed in the visually striking lateral arch, giving an impression of robust, lightweight strength.

Concept Car Racoon
by Renault

The name for Racoon is taken from Northern American mammals so well known for its love of climbing, water and cleanliness.

373

Bunny Courses by Design (Pool)

Nissan Forklift by Achilles Associates bvba

Venet Sea
by Design (Pool)

Bova Magiq
by Enthoven Associates

**Bombardier M6 doubledeck
trainseats**
by Enthoven Associates

Amtrak High Speed Train
by Henry Dreyfuss Associates

Pendo *by Priestman Goode*

Airbus A3XX Bar *by Priestman Goode*

Airbus A3XX *by Priestman Goode*

Virgin Train *by Priestman Goode*

American Airlines 767 interior
by Henry Dreyfuss Associates

Airbus A3XX Library *by Priestman Goode*

Public

Public transport is increasingly forging its own identity and in the process it is breaking ground with respect to both safety and passenger comfort.

Öffentliche Verkehrsmittel

Die öffentlichen Verkehrsmittel gewinnen zunehmend an Bedeutung, wobei sowohl die Sicherheit der Fahrgäste als auch die Bequemlichkeit und der Komfort laufend verbessert werden.

Publics

Le transport public acquière chaque jour une importance plus spécifique et, ainsi, engendre de nouvelles innovations tant en matière de sécurité comme pour le confort du passager.

Públicos

El transporte público va tomando cada vez más peso específico, de forma que se están logrando innovaciones tanto en materia de seguridad como en el ámbito del confort del pasajero.

Alstom – Ret Rotterdam *by Enthoven Associates*

Bombardier Amsterdam concept study *by Enthoven Associates*

Bombardier extensions
Belgian cost-low floor extensions
by Enthoven Associates

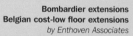

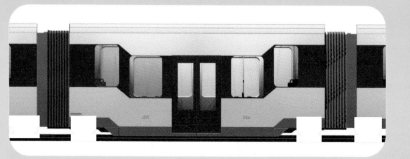

Long Island Railroad
by Henry Dreyfuss Associates

Alstom Light Rail Concept
by Henry Dreyfuss Associates

Suburban Train
for Renfe by Mormedi

Virgin Atlantic Upper Class Seat Front
by Priestman Goode

Virgin Atlantic Upper Class Seat Bed
by Priestman Goode

WCML Premium Saloon
by Priestman Goode

Think *for Ford Motor Company by Frog Design*

A revolutionary new electric vehicle. A different driving experience with a innovative design.

Seats *for Lufthansa by Frog Design*

Triumph *by Ideo Europe*

IVM *for BMW by Mormedi*

*This car packs sporty
performance and celebrates
a return of the "pure joy of
driving".*

Concept Car Laguna
by Renault

Rinspeed Advantige Rone
by Rinspeed Design

Innovation backed by tradition-creation of a strikingly designed, down-to-earth sports car that offers unique driving fun.

Private
This is a design field with a great symbolic power, in which every company seeks a corporate identity with an emotional impact to distinguish it from the competition.

Privatbereich
Dieser Sektor ist mit großer Symbolik verbunden, und jede Firma ist auf der Suche nach einer emotionsgeladene Corporate Identity, die sie von den anderen Fabrikanten unterscheidet.

Privés
C'est un champ du design affecté d'une grande charge symbolique, où chaque entreprise recherche son identité, chargée d'émotions qui la différencient des autres sociétés.

Privado
Es un campo del diseño con una gran carga simbólica, donde cada empresa busca una identidad corporativa cargada de emociones que la diferencie de las otras casas.

Rinspeed Tatoo.com *by Rinspeed Design*

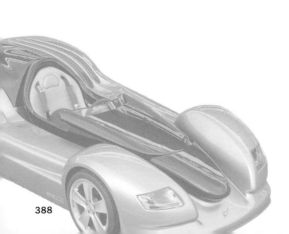

Rinspeed X-Trem *by Rinspeed Design*

Rinspeed X-Tra-Lift
for Ford F250 by Rinspeed Design

Concept Car Scenic *by Renault*

Concept Car Mégane *by Renault*

Concept Car Scenic interior *by Renault*

Concept Car Talisman *by Renault*

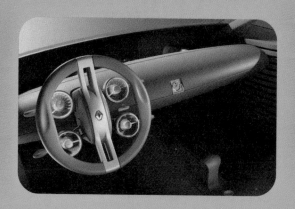

Concept Car Argos *by Renault*

Rinspeed Presto
by Rinspeed Design

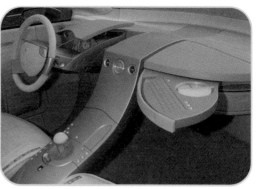

Concept Car Zo
by Renault

Concept Car Vel Satis *by Renault*

Concept Car Scenic *by Renault*

Renault 28 *by Renault*

Concept Car Fiftie *by Renault*

Concept Car Pangea *by Renault*

Concept Car Operandi *by Renault*

Directory

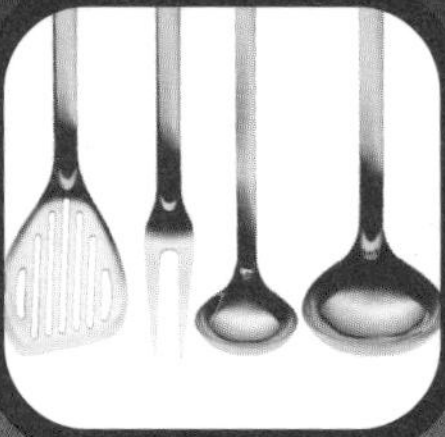

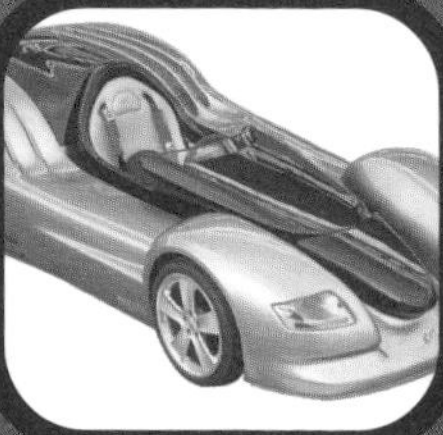

Verzeichnis

Table des Matières

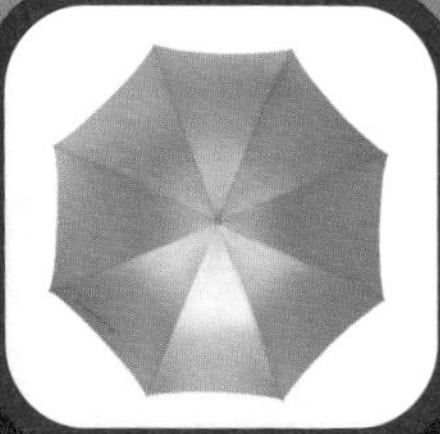

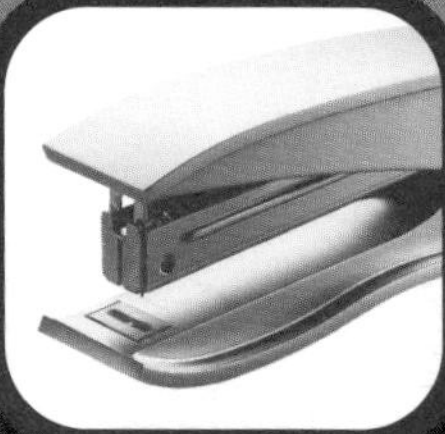

Directorio

Other Designpocket titles by teNeues:

Berlin Apartments 3-8238-5596-4

Cafés & Restaurants 3-8238-5478-X

Cool Hotels 3-8238-5556-5

Country Hotels 3-8238-5574-3

Exhibition Design 3-8238-5548-4

Furniture/Möbel/Meubles/Mobile Design 3-8238-5575-1

Italian Interior Design 3-8238-5495-X

London Apartments 3-8238-5558-1

Los Angeles Houses 3-8238-5594-8

New York Apartments 3-8238-5557-3

Office Design 3-8238-5578-6

Paris Apartments 3-8238-5571-9

Showrooms 3-8238-5496-8

Spa & Wellness Hotels 3-8238-5595-6

Staircases 3-8238-5572-7

Tokyo Houses 3-8238-5573-5

Each volume:

12.5 x 18.5 cm
400 pages
c. 400 color illustrations